EXTREME WAVES

CRAIG B. SMITH

ILLUSTRATIONS BY KURT MUELLER

Joseph Henry Press
Washington, D.C.

Joseph Henry Press • 500 Fifth Street, NW • Washington, DC 20001

The Joseph Henry Press, an imprint of the National Academies Press, was created with the goal of making books on science, technology, and health more widely available to professionals and the public. Joseph Henry was one of the founders of the National Academy of Sciences and a leader in early American science.

Any opinions, findings, conclusions, or recommendations expressed in this volume are those of the author and do not necessarily reflect the views of the National Academy of Sciences or its affiliated institutions.

Library of Congress Cataloging-in-Publication Data

Smith, Craig B.
  Extreme waves / by Craig B. Smith ; illustrations by Kurt Mueller.
    p. cm.
  Includes bibliographical references and index.
  ISBN 0-309-10062-3 (cloth)
  1. Ocean waves. I. Title.
  GC211.2.S65 2006
  551.46'3—dc22

                                    2006019554

Cover design by Michele de la Menardiere

Cover illustration © W. Faidley/Weatherstock Photography

Printed in the United States of America

*This book is dedicated to those
who share my love of oceans.*

Contents

Foreword   xi
   Dr. Susanne Lehner
Preface   xvii
Introduction   1
  1 The Calm Sea   11
  2 The Four Winds and Waves   25
  3 Over the Bounding Main   47
  4 Tempests and Storm-Tossed Seas   71
  5 Swell   110
  6 Terror Waves: Tsunami   127
  7 The Southeast Asia Tsunami of December 26, 2004   148
  8 A Confused Sea   164
  9 Freaks, Rogues, and Giants   184
 10 When the Big Wave Comes: Are Ships Safe Enough?   216
 11 Davy Jones's Locker   226

Appendixes
A Recent Research on Extreme Wave Models   247
B Units of Measure and Conversion Factors   251
C Glossary of Special Terms   253

vii

Endnotes	256
Annotated Bibliography	269
Permissions and Credits	275
Acknowledgments	277
Index	281

Plates, Figures, and Tables

PLATES

1   Dan Moore surfs a 68-foot-high wave
2   Tsunami damage at Hilo, Hawaii, May 22, 1960
3   *Bobsled* in the 1998 Sydney-Hobart race
4   Crew prepare to abandon dismasted and sinking *Stand Aside*
5   (A) A brisk sail; (B) Wind-blown wave in a storm
6   Tidal bore, Qiantang River, China
7   The wreck of the *Memphis* in Santo Domingo (1916)
8   U.S. Navy Indian Ocean wave forecast
9   Tsunami flooding at Laie Point, Hawaii, March 9, 1957
10 Man caught in tsunami, Phuket, Thailand, December 26, 2004
11 Tsunami-damaged bungalow, Koh Phi Phi Island, December 26, 2005
12 Wave victim on the Agulhas Coast, South Africa
13 Freighter in a typhoon, North Pacific
14 Merchant vessel *Winter Water* fights heavy seas from tropical storm *Georgette*, Eastern Pacific Ocean, July 28, 1980
15 North Pacific Ocean storm waves as seen from M/V *Nobel Star*, winter, 1989

FIGURES

1 Types of ocean waves, 13
2 Properties of an ideal wave, 14
3 Ocean surface winds in August, 33
4 Major ocean currents, 36
5 How a barometer works, 43
6 Destructive (A) and constructive (B) interference of waves, 50-51
7 A composite wave with seven components, 52
8 Rayleigh distribution of wave heights, 55
9 Typical NOAA marine weather wind-wave forecast, 57
10 Global significant wave heights, H_S, in February (in meters), 58
11 Wave steepness, 64
12 Wind-wave patterns from a hurricane, 98
13 Tsunami wave height and run-up, 129
14 Map of Thira Island and vicinity, 131
15 Tsunami travel time in hours, Chile earthquake, May 22, 1960, 136
16 Map of Koh Phi Phi Island and vicinity, 153
17 Wave diffraction, Newport Harbor entrance, 170
18 Draupner oil platform wave height, January 1, 1995, 208
19 USS *Ramapo* wave height measurement scheme, 213

TABLES

1 The Earth's Oceans and Seas, 16
2 Wave Height Versus Wind Speed, Duration, and Fetch, 41
3 Deadliest Hurricanes, Continental USA, 1900-2004, 100
4 Worst Hurricanes, 109
5 The Evidence for Extreme Waves, 215
6 Distribution of Vessels in the World Merchant Fleet, 217

Foreword

In 2002, while we were doing research on rogue waves in the MAXWAVE project, a sudden spurt of interest from the media swept the globe. Incredible! The web page of the European Space Agency— showing images of rogue waves taken by a radar satellite at night and through the clouds—recorded more hits than some exciting recent news about film stars. E-mails, phone calls, and requests from TV teams poured in; our scientists—especially those with sailor's beards—were in high demand as interview partners. It is fascinating that we are now able to observe from space the most dangerous waves roaring across the oceans and even do this in the comfort of an air-conditioned computer room. Now satellite measurements can be acquired in remote areas that were not observed before.

From the time of the earliest civilizations, humankind has been fascinated with stories of giant waves—the "monsters" of the seas. So a book on rogue waves is obviously of interest to a wide range of people, from surfers to those who crisscross the oceans in the interest of commerce or peace and national defense, and even cruise ships, which today follow routes that in the past only adventurers would have dared to explore. And while scientists continue to debate the details of the nature and physics of such waves, this book provides a fascinating, highly readable account of what we know so far about rogue waves.

Extreme ocean waves are one of our most frightening images—towers of water pounding down on a helpless boat. You can observe the wall of water coming; you can even calculate when it will arrive, but you cannot run away and you cannot fight it—a classic nightmare. Can we cope with it in the future? Predict extreme waves? Control them? Ride giant waves like surfers?

A tragic realization of this nightmare was the December 26, 2004, Southeast Asia tsunami, which was triggered by an earthquake beneath the Indian Ocean. Tsunami are the result of earthquakes that cause large bodies of water to be shifted, creating a long, low wave in the open ocean—but one that rises up and causes tremendous destruction when it approaches shore. Rogue waves on the open ocean—the waves that sink ships and damage offshore structures—are caused by storms.

I have received many interesting e-mails offering explanations as to how rogue waves are formed. One individual suggested the possibility that a huge wave is reflected by a cliff and then by chance combines with another such wave that is coming from the other side of the ocean. Another thought that these waves might be generated by little black holes, suddenly springing into existence and causing disruptions in gravity that create walls of water.

The media who contacted us were most interested in hearing about monster waves—immense waves that appear suddenly, swallow a ship, and then vanish as quickly as they arise. There is a scientific name for consecutive rogue waves to which ships fall victim—"the Three Sisters"—a term applied to a sequence of three waves that are considered to be as dangerous and unpredictable as three women aiming for the same parking spot at a crowded mall. I suspect, however, that the real explanation of why ships founder is that more accidents happen when a ship's crew is challenged the third time.

In any case, I have not found such patterns in the satellite images I scan for rogue waves. Such simple explanations as high waves being generated in extended storms or the superposition of crossing waves from two different storms seem more reasonable.

Sailors, obviously, are the ones who have frequent encounters with extreme waves. Races such as the Fastnet Rock and the 1998 Sydney-Hobart made quite an impression on the sailing community because of the rogue waves participants encountered—leading to new descriptions and fascinating photos of the waves. The harsh reality, however, is that when confronting these terrifying waves, many otherwise courageous sailors simply panic, abandoning their vessels—or, worse, their comrades.

History teaches us, of course, that venturing out to sea can prove dangerous. We accept the fact that some vessels will not return home safely—an acceptance that we would find unthinkable for air travel. The shipping community is conservative, not easily open to changes. There are no black boxes on ships; no recorded last messages of the captain. Detailed results of investigations—why a ship sank, why the windows broke or the hatch locks on deck cracked, why the containers moved or the engines stalled—are not present in the news, that is, if the incident makes the news. These tragedies often are attributed to severe weather, something that is unavoidable from time to time. It seems that the main interest in ship accidents still focuses on passenger ships; maybe things have not changed that much since the *Titanic*, with first-class people coming first? Everything else insured?

Thanks to advances in satellite technology, extended in situ measurements, and computer capabilities, vessels can improve the odds of avoiding an accident at sea due to severe weather. Statistics of past events help us predict the so-called 100-year waves in various areas of the ocean and provide a better understanding of the conditions under which ships' crews must operate. Engineers designing ships and building offshore platforms that are farther and farther from land obviously require detailed knowledge of structural loads imposed by wind and wave.

Satellite images of the North Sea do not show an empty space, but rather an area dotted by bright spots of oil platforms more densely inhabited than many land areas. The offshore industry is propelling the advancement of research on rogue waves. There are significant financial ramifications if production at offshore facilities must be halted or platforms evacuated because of the danger posed by a rogue wave. For this reason, many oil platforms are equipped with ocean wave measurement instruments such as radar and laser altimeters and marine radar; buoys are deployed nearby and satellites are watching from above. Still, recent observations of high waves could not be explained by current models. And because platforms cannot move away like a ship, there is an additional incentive to improve construction, especially as the deck sinks during the years of extracting the oil underneath. Engineers are practical people; they have to build structures that withstand such waves.

It is lucky if a rogue wave is even recorded. Often the measurements are ignored because people do not believe in them, or they are not stored but are classified as "noise" (spurious data) straight away.

What is the highest wave ever observed? Has anybody validated the satellite measurements by in situ observations? Has somebody actually measured a 40-meter-high wave somewhere in the Antarctic winter? Or is it all sailors' tales? Craig Smith told me that besides being a sailor, he has spent time on an offshore platform that had been hit by a giant wave. So, this book brings realism to the subject for the general reader.

Can we predict rogue waves? At the moment, weather centers usually issue forecasts based on the significant wave height. Scientific discussions about the causes of extreme waves are still going on, so proposals for new research or reviews of scientific articles may get top marks from one reviewer and complete rejection from another.

Clearly, extreme waves need much more study. For, as Craig Smith explains, numerous lives have been lost and will continue to be lost because ships cannot withstand the forces they unleash. It is our fervent hope that the shipping community will be able to make use of the data gathered by scientists to make ships and offshore platforms more robust. I myself plan to observe the behavior of surf by radar measurements (preferably in Hawaii); to go on a scientific expedition near Antarctica, where the highest waves seem to occur; to validate satellite measurements of such waves from a research vessel; to generate a wave atlas of a global 10-year time series of satellite measurements of extreme wave heights;, to complete some written work on this topic as well; and hopefully to find some time to read books such as this one.

Prof. Susanne H. Lehner
Rosenstiel School of Marine and Atmospheric Science
University of Miami
and
Remote Sensing Technology Institute
German Aerospace Center
Oberpfaffenhofen

Preface

As a boy growing up in Southern California, I was never far from the Pacific Ocean and my family spent time on the local beaches during the summers. I recall lazy afternoons bodysurfing at Huntington Beach or going to Balboa Island and swimming in Newport Harbor. I had my first sailing experiences in the harbor with high school friends. We were all avid swimmers, and we went skin diving and spear fishing at nearby Laguna Beach. Turning our backs on land, looking out to sea, all noise and distraction were left behind. The ocean shuts out all other sounds. At night we built a fire and cooked our catch on long empty beaches that today are hemmed in by hotels and exclusive beachfront homes. Other memories are of fishing trips to Catalina on charter boats with my father and his friends. From these experiences came a love of the sea that has never diminished.

At the time our daughter Kelly was born in 1968, my wife, Nancy, and I moved with our three-year-old son Kent and new baby to Pacific Palisades, California, to house-sit a friend's home. The house was an old Spanish-style three-story structure on a hillside high above the Pacific Coast highway. A walkway with a thousand or more steps led down the hillside to a pedestrian bridge over the highway and onto Will Rogers State Beach where we could swim, picnic, or surf-fish.

From the balconies on three sides of the house it was possible to look out over the ocean to the south across Santa Monica Bay to Palos Verdes Point and to Catalina Island. On a clear day you could see Santa Barbara Island to the southwest and, to the west, Point Dume and the Channel Islands somewhere beyond. On winter nights when the swell from distant storms reached the coast, you could hear the waves crashing on the beach below. In the late afternoon, the sun illuminated the corrugated appearance of the sea stretching out to the horizon, the rippling water shining in an infinite variety of sparkling reflections and waveforms until at sunset as the sun disappeared over the horizon the colors would turn from blue to orange to gray. On clear nights with a full moon, a long shaft of yellow moonlight stretched out across the sea, the waves clearly visible as they made their way to shore. At night on a balcony, watching the waves, I wondered about their origins—where they came from, why they behaved the way they did, what brought about the variations in their behavior.

Finally a day came when our friends returned, and we had to exchange our wonderful beach house for one in the city—a "starter home" that became our residence for the next 30 years. We were only 15 or 20 minutes away from the beaches, but it was not the same. Fifteen years after moving from the beach house, I bought a half interest in a 26 foot-long-sloop named *Karess* and could once more experience the infinite variations of ocean winds and waves, this time from the deck of a sailing vessel. At sea in a small boat you cannot help but take a personal interest in wind and wave and how they can toss a boat.

I was familiar with famous shipwrecks of past centuries—the loss of Spanish galleons and treasure fleets during Caribbean storms or the collision of the *Titanic* with an iceberg while on its maiden voyage. You would think that with the technologies available during the last two centuries, the rate of loss of ships and human lives would have dropped dramatically. Surely radar, improved design, periodic safety inspections, construction using high-strength steel, and satellite weather observations would ensure that ships are lost much less frequently today.

However, in the course of conducting research for this book I learned that the frequency of ship disasters remains surprisingly high.

Where do waves come from, and why are some small, some large? In Southern California where I live, every surfer knows that waves come in sets, and it is—depending on the day and conditions—that seventh or tenth wave that promises a long exhilarating ride to the beach. At my present home on the Balboa Peninsula, a few hundred yards from the ocean, I can hear the surf at night. On some nights it is a distant whisper; on others it crashes and rumbles, changing with the tides and distant storms. Let a hurricane pound Mexico's Baja California 1,200 miles away, and a few days later surfers flock to Newport Beach to take advantage of the southerly swell that arrives, creating waves that crest from a few to a dozen feet in height. There have been occasions when I've sat on beaches in California or Hawaii completely awed by the majestic power of storm waves so immense and so violent that I would not have dared to enter the water. (See Plate 1.)

Others are bolder than I. Roughly 87 nautical miles from the California coast lies Cortes Bank and Bishop Rock, the latter named for a ship that foundered there years ago. I've visited Cortes Bank in my sailboat when the swell was less than 3 feet high. With the boat standing off a few hundred yards from the bank, you can see the swell approach and then rise up in long, glistening waves as the ocean goes from extreme depth to less than 5 fathoms or 30 feet over the bank. Following winter storms, larger swells approach the bank and generate waves 50 to 65 feet high. Surfers have been known to ride these giant waves, traveling to the bank by boat and then being towed out by wave runners at 40 miles per hour to catch the wave.

I've experienced rough seas in a 50-foot fishing boat making night crossings of the Kauai Channel between Oahu and Kauai or the Kaiwi Channel between Oahu and Molokai, and I've sailed my boat among California's offshore islands when small craft warnings were posted. These experiences suffice; from them I learned the taste of an angry sea. My mouth was metallic dry with fear—even though I was not worried about the sturdy vessel I was in—but you never forget the screaming wind in the standing rigging as it vibrates under the force of mast and sail. There was no escaping the knowledge that I was captive to forces more powerful than me and that it required my skill and a seaworthy craft to sail into the gray sky and changing landscape of a foaming, stormy sea.

I write these pages in my friend Andy Youngquist's boat, *Sitting Tall*, a 46-foot Bertram. It is a superb sports fisher, powered by two Detroit diesel engines capable of pushing it to 20 knots. At this early hour, just before sunrise, we are on a mooring in Avalon Harbor, Catalina Island, rocking to a gentle swell. Quite different from yesterday, when a Santa Ana wind came up suddenly on the mainland around 26 nautical miles away, and shortly after we'd left the harbor sent waves 3 to 6 feet high into the harbor, bouncing floating docks and causing one boat to founder and sink. This was a perfect example of the interaction of wind and sea; the average height of the waves is determined by the speed of the wind, the length of time it blows, and the *fetch*, or distance of open water, over which it blows. These were not large waves, but in the tight confines of a small harbor, they had the potential to cause a great deal of mischief.

So, the sea reminds me—there are small waves and there are very large waves. As you can see from this book, giant waves—waves so large they tower as high as a 10-story building—are more common than previously believed. Most of the merchant and passengers ships that ply world trade routes today are likely to come out second-best in an encounter with an extreme wave of this height. Along the coasts of the oceans, tsunami are another source of high and destructive waves. My purpose in writing this book is to increase awareness of the hazard represented by giant waves and to suggest ways to mitigate the risk for those of us who live near the sea or whose occupations take us across oceans.

May you have fair winds and following seas.

Craig B. Smith
Balboa, California

Introduction

Since you are reading this book, there is a good chance that at some time you have found yourself sitting on a beach, watching the surf—the endless progression of waves crashing on the shore. Or perhaps you have memories of walking along the shore, dancing back to keep your shoes dry as a wave expends itself, rushing up the beach. Fortunately, few of us on shore or in a boat have ever looked out to sea and seen a wave as tall as a 10-story building racing toward us, and in that instant known that there was no way to outrun it, no way to survive, and that our life was about to come to an end.

Impossible? Implausible? How about a wave so high that it would wash up a mountainside higher than a 100-story building, to rip boulders and trees from the ground? Can you imagine being in a small boat and seeing *that* wave bearing down on you? On July 9, 1958, three boats were hit by such a wave in Lituya Bay, Alaska. Two of the boats sank, but the other one—miraculously carried out and then back into the bay by the wave—survived. There were eyewitnesses who saw this wave. They estimated that while riding the wave their boat cleared a spit of land at the entrance to the bay by 100 feet or more. The maximum wave run-up, or height reached on dry land, is not an estimate; it is known *exactly*. The wave stripped all the trees off the mountains ringing the entrance of the bay to an elevation of 1,700 feet. This is not the largest wave ever recorded, although it is the largest wave that eyewitnesses are known to have survived. There is geologic evidence of even larger waves in prehistoric times—some caused by earthquakes, others presumed to have been caused by the impact of a meteorite in the ocean.

On average, several major vessels are lost every single week somewhere in the world, many from the effects of extreme waves. Not too long ago, a major vessel was lost every day of the year. This statistic does not consider the losses of fishing boats, small boats, and pleasure craft. Peacetime ship losses are reported in six major categories: wrecked; burned; collided; foundered; missing, presumed lost; and "other."[1]

"Wrecked" is the most common cause of ship loss; the term means run aground, as the result of either foul weather, equipment failure, or human error—usually navigational error. Fire or explosion is another significant cause of ship loss, due both to the cargo that ships carry and the fuel used for propulsion. Even with radar, traffic separation lanes, and radio communications, collisions still occur—in part perhaps because of the larger, faster vessels in use today—vessels that may require a mile or more to stop or change course. "Other" includes miscellaneous causes; amazingly, even today piracy accounts for ship losses.

The two remaining categories are important for this book. *Foundered* means sunk at sea—overwhelmed by the influence of wind and wave. Vessels founder because of heavy storms, although human error (mishandling of the ship) or equipment failure (engines fail; hatches or other openings admit water into the vessel) may also contribute to a ship's sinking. Ships also founder as a result of extreme waves—either a single large wave or several waves in quick succession—that simply overwhelm the vessel, rolling it, damaging it, or in some cases breaking it apart before the crew has time to take evasive action or even make a Mayday call on the radio.[2] A wave need not be hundreds of feet high to be considered "extreme." A wave that is 66 to 98 feet high is large enough to pose a dire threat to large vessels at sea or to structures along the coast.[3]

The last category—missing and presumed lost—is in my mind the most dreadful. Ships listed as *missing, presumed lost* are exactly that; they have suddenly disappeared, no trace of them has been found, and no reason for their sudden disappearance can be given. There are two probable causes for a ship to suddenly disappear without a radio Mayday message, without launching an emergency position-indicating radio beacon, and without any survivors: a massive explosion that breaks up the vessel and causes it to sink immediately or a massive wave that overwhelms the vessel, capsizes it, and drives it and its unfortunate crew to the bottom of the ocean. Every year a number of vessels end up in the missing category. The families of crew members and the vessel owners then have to live with the agony of never knowing what really happened to loved ones, cargo, and property. Once "missing" appears on Lloyd's Casualty List, claims can be filed and relatives can begin the mourning process.[4]

History tells us that in the five millennia during which humans have ventured into the sea, many have lost their lives, and ship sinkings or shipwrecks have not been uncommon. On the basis of British records for the 1700s and 1800s, noted oceanographer Willard Bascom estimated that 300,000 ships were wrecked per century. Another 100,000 per century foundered—that is, they sank in some distant ocean, away from land, usually with the loss of the entire crew.[5]

Bascom states that for the 10-year period 1960-1970, 2,766 major ships—500 tons and larger—were lost, an average of 277 per year. Of these, 1,136 (41 percent) were wrecked. The next largest amount, 771 (28 percent), foundered and 70 (about 2.5 percent) were declared missing. The basic data come from Lloyds and other maritime insurers. In 1980, 387 vessels were lost from a world fleet of 73,882.[6]

This general trend continues today. For example, for cargo vessels only—tankers, bulk carriers, and container ships—during the winters of 1997-1998 and 2004-2005, 80 and 30 vessels, respectively, were lost in a 121-day period.[7] In 1997-1998 the causes and number were foundered (26), collision (22), wrecked (17), explosion/fire (14), and missing and presumed lost (1). Of those vessels that foundered, 20 incidents were caused by heavy weather and storms. In 2004-2005, the corresponding numbers were foundered (14), collision (3), wrecked (12), and explosion or fire (1). The brief reports I examined were insufficient to identify rogue waves as a specific cause, although they were indicated in several cases. For example, the container ship M/V *MSC Carla* broke in two during a storm in the North Atlantic, and one-half (with its 1,200 containers) sank quickly, while the other half stayed afloat and attempts were made to take it in tow and salvage it. There were two other incidents that reported a total of 86 containers lost overboard. Some of the wrecks were caused by navigational errors, but most were due to storms and foul weather. Eight of the 17 wrecks were ships grounded on Guam by Typhoon Paka in December 1999. These numbers do not include fishing boats, military vessels, passenger ships, or ferries. If commercial fishing vessels were included, the losses would double at least.

The number of ships lost has decreased somewhat in the last two decades, but this is misleading, since the size of the world merchant fleet has changed. In 1980 there were 73,882 vessels, versus 39,932 in 2005. Today, the average vessel is much larger. From 1992 to 2003, a total of 1,049 merchant vessels were lost. During this time, bulk carriers (cargo vessels designed to carry grain, cement, iron ore, or similar bulk commodities) represented the greatest number of ships lost every year except 1997 and 2003. Roughly speaking, a bulk carrier or tanker was *lost every other week.* In 2003, 91 cargo vessels (bulk carriers, tankers, container ships, and others) exceeding 500 gross tons were lost. During 1992-2003, foundering was the major cause—30.9 percent—of all losses.[8]

Looking at vessel types, from 1974 to 1988, between two and three tankers per hundred suffered a serious casualty, usually resulting in an oil spill. Given that there were 5,000 to 10,000 tankers in service at these times, hundreds of vessels suffered serious casualties each year.[9] During the 1992-2003 period, most of the tanker losses were caused by fire and explosion, while the majority of the bulk carrier losses (42 percent) were due to weather conditions that caused the vessel to founder or run aground.

Since several major ships are lost every week and around 30 percent founder in heavy weather, improved design to resist extreme waves and tactics to avoid encountering them could have a measurable impact on maritime safety. Improved safety requires a better understanding of how giant waves arise.

Waves change with the seasons, every year reshaping the beaches they impinge. From winter to summer the nature of a beach changes, sculpted by the tireless energy of the waves. In the winter, my beach develops a sharp drop-off, 3 to 8 feet high, as distant winter storms send breakers that carry sand out to sea. Later, other waves return the sand, and the beach returns to its characteristic slope to the sea. Usually, these changes are gradual, but sometimes, under the combined forces of a storm and an unusually high tide, the contours of the beach change dramatically overnight.

The sculpting touch of the waves reveals hidden treasures. On one day, the sand might be swept bare; on the next, it might reveal a pattern of shells drawing a wavy line previously hidden from sight. Or wind and wave may combine to strand sea creatures on the shore. Once, *Velella velella* (by-the-wind sailor, a type of small jellyfish) appeared by the thousands along the length of my beach, their mysterious journey to find mates and to reproduce interrupted by a chance wind that blew them into the surf and left them stranded on the beach.

Other sea animals know how to use the waves to their advantage, none being better adapted to the surf than the small fish called grunion. During the spring and summer months, when the grunion spawn, they wait for high tide. On several nights following the highest tide (the third, fourth, or fifth night following a full or new moon), they allow a wave to strand them on the beach, where the female scoops a shallow depression in the sand and deposits her eggs. Nearby males fertilize the eggs and then both fish wriggle their way back to the water. The eggs mature in about 10 days, but do not hatch until the next high tide washes them free of the sand and enables the hatchlings to make their way to the sea.

Watching waves from the security of a sandy beach is one thing; watching them from the deck of a pitching ship is another. My intimate knowledge of waves comes from observations from the decks and cockpits of both small and large vessels. I've crossed the North Atlantic in December on the SS *United States*, 990 feet long, capacity about 2,000 passengers, which at the time was the fastest cruise ship ever built. It averaged 35.6 knots when it set the record for the crossing from Southampton to New York. I've also traversed the Mediterranean north to south on another cruise ship and have been in the Pacific on *Cygnus Voyager*, a million-barrel-capacity crude oil tanker. However, I've never ventured to the Southern Ocean, nor do I desire to do so.

Seafarers' lore is replete with tales of ships mysteriously disappearing or being battered by terrific wind and waves in fierce storms. It is well known that it is difficult to judge the height of a large wave accurately from the deck of a pitching ship, simply because there is no point of reference, nothing to which the eye can compare it. When eyewitness accounts can be compared to more accurate measurements, it is found that the waves are always smaller than they seem. For this reason, tales of waves 100 feet high have been discounted in the past, except in those few cases in which some type of comparative measurement has been possible.

Extreme waves can arise from several different sources—tropical cyclones and storms being the most obvious cause, but possibly not the most common nor the most deadly. Large, fast-moving waves, called *tsunami*, can arise from undersea earthquakes or even massive landslides or the collapse of a large ice wall from a glacier. A more remote source is a meteorite striking the ocean, an event suspected to have occurred in ancient times with disastrous results. Finally, in the complex interactions that take place in the sea during which various waves from sources near and distant eventually come together—in some cases canceling each other and in other cases reinforcing each other—suddenly a wave much larger than the rest will appear. Some giant waves will disappear as quickly as they appear; others, fed by the energy of a storm, will continue to travel and grow. In either case, unlucky is the hapless vessel that finds itself in the path of such a wave.

Maritime history abounds with tales of ships being lost in mysterious ways—no survivors to describe what happened, or those who survive recounting a version of events bordering on the fantastic. Extreme waves have been reported from ancient times and are mentioned by Homer (eighth century B.C.) in *The Odyssey*. Odysseus, after offending

Poseidon, god of the seas, sees his raft destroyed by a mighty wave. In the 1500s, at the height of the Spanish conquest of South America, dozens of square-rigged galleons were lost in the Gulf of Mexico and the Gulf Stream off Florida, victims of hurricanes and extreme waves that suddenly overtook the cumbersome vessels and sent them to the bottom.

In more recent times, the saga of the paddle wheel steamer *Central America* typifies the sinkings in those waters. Known as the "Ship of Gold," the *Central America* was traveling from Panama to New York with almost 600 passengers and crew, many of them miners returning from the California gold fields. The ship transported tons of gold in the form of ingots, nuggets, gold coins, and gold dust. A number of the returning miners had made their fortunes in the gold rush and were returning east to resume less arduous lives. A week out of Panama, the ship encountered a gale. Fierce winds shredded the sails, and waves damaged the paddle wheels and steering. Water began to fill the hold, flooding and extinguishing the boilers. With no power and steerage, the vessel rolled helplessly in mammoth seas.[10]

When all appeared lost, another vessel—the brig *Marine*—appeared on the horizon. The *Central America* was able to evacuate women and children to the *Marine*, but further damage to both vessels precluded the removal of the remaining passengers. On the night of September 12, 1857, the vessel started to sink. In the horror and fear precipitated by the creaking and groaning of the dying vessel settling into the sea, one survivor recalled:

> The love of gold was forgotten. Men unbuckled their gold-stuffed belts and flung their hard-earned treasure on deck, to lighten their weight. Anyone could have had a fortune, just for picking it up, but with no chance of reaching safety with this treasure.[11]

The *Central America* sank in 1,312 fathoms of water, about 173 nautical miles east of the Carolinas. Of the passengers left behind, another 50 or so were rescued, but a total of 425 died.

This story has a remarkable ending. Almost to the day, 131 years later, on September 11, undersea explorer Tommy Thompson finally found "the Ship of Gold" after a search of several years. Thousands of gold coins, ingots, nuggets, and even gold dust have been recovered, along with a treasure trove of clothing and other artifacts that reveal what life in the gold fields was like.

The so-called Bermuda Triangle, an area delineated by Miami, Florida; San Juan, Puerto Rico; and Bermuda, is the source of legends of mysterious disappearances of aircraft and ships. More likely, sinkings are the result of sudden weather changes that occur in the region. Storms, in combination with the strong current of the Gulf Stream, as well as sea bottom variations that range from shoals around some of the islands to some of the deepest marine trenches in the world, can combine to create large waves.

The random occurrence of extreme waves—especially when the sea is not particularly rough—was previously thought to be very rare. Recent research, however, is revealing that giant waves (or rogue waves, as they are sometimes called) are more common than previously thought and are probably behind the disappearance of many vessels that have been mysteriously lost at sea. Giant waves are all the more frightening because by their very nature they can appear out of a seemingly calm sea, posing a serious risk even to supercarriers or supertankers.

Some years ago I spent several weeks at the Ekofisk North Sea oil field, conducting structural tests on an offshore platform named "Two-Four-Delta." Ekofisk is an area of the North Sea where the water depth is around 100 feet and the platform legs extend 100 feet into the seafloor. The "100-year wave"—the wave so large that it will come only once in 100 years—is the design criterion for these structures. The 100-year wave for Two-Four-Delta was 100 feet and actually came in year 2, as the platform was being constructed. Since all of the superstructure had not been installed, the platform survived with minor damage. On my visit, I spent some time on the "spider deck," basically a series of catwalks about 30 feet above the sea. There I could observe massive steel I-beams, bent and twisted by the force of waves.

It is not just ships at sea that have to be concerned with extreme waves—ports, harbors, offshore structures, coastal cities, and seaside resorts can also feel the wrath of the sea when natural forces of weather and geology become extreme. The recent December 26, 2004, magnitude 9.0 Sumatra-Andaman Islands earthquake and the resulting tsunami provided the latest and one of the most terrifying examples of the power of an aroused sea when pitted against human habitation. Over the centuries, tsunami have devastated coastal areas of Japan and China, where some of the oldest written records are to be found. Also, in more recent times, the destruction of huge swaths of coastal Chile and Peru are grim reminders of the power of the oceans when unleashed as huge waves.

Two images always come to my mind when I think of tsunami. The first is Hilo, Hawaii, where photographs taken after the tsunami showed rows of parking meters bent over and lying parallel to the street, a silent symbol of the force of the wave that bore down on them. (See Plate 2.) The second is U.S. Coast Guard photographs of the Scotch Cap lighthouse on Unimak Island, Alaska. In the "before" photograph, the lighthouse stands tall, its beam 92 feet above the level of the sea, a rugged structure of steel-reinforced concrete. In the "after" photograph, taken a few days after a tsunami hit the lighthouse, the bluff on which the lighthouse stood is littered with debris. The lighthouse is gone, totally erased from the landscape. Also erased were the lives of the five crew members who stood watch on that fateful night.

So the sea—our friend, our benefactor, vital for food, essential for transportation, critical for controlling the earth's climate—is also a source of danger. Who would have thought that thousands of international tourists, vacationing on the sunny beaches of Indonesia and Thailand, would have but minutes to act if they were to save their lives on December 26, 2004? This frightful scenario could reoccur on other beaches at some future date. What about the millions of people living on the fringes of the world's great oceans? With improved warning systems and education concerning tsunami dangers and evacuation procedures, many lives would be saved. And if it is true that extreme waves—waves capable of breaking and sinking major ships—are more likely to occur than previously believed, ship design standards need to be revised. These are reasons enough for everyone to understand how extreme waves are formed and propagate, and how their risks may best be managed by improved forecasting, improved warning systems, and improved design.

1

The Calm Sea

Once, as I was sailing from Marina Del Rey, California, to Baja California, Mexico, the fickle morning wind slowly diminished and disappeared by noon. The sea took on a glassy sheen, marred by nothing but the faint trace of a southerly current. The sails hung listlessly, barely moving with the slight roll of the boat. *Dreams*, my 33-foot-long, cutter-rigged sailboat, was becalmed.[1] My charts put us a little south and east of Sixty Mile Bank, an area of submarine mountain ranges and deep canyons—a favored destination for fishing boats operating out of San Diego. It was a warm day, and after trailing a line astern for the safety of the swimmers, some of the crew went overboard for a refreshing swim in midocean, around 54 nautical miles from the nearest land, then returned to await the coming of the afternoon breeze.

As flat as the sea was that memorable day, there is in fact no such thing as a perfectly calm sea. In the absence of any other disturbing force, planetary gravitational forces are constantly at work on the earth's oceans. Newton's law of universal gravitation tells us that there is a mutual attraction between two bodies. In the case of the earth, the situation is more complicated, since it involves forces between the earth, sun, and moon; and even more distant planets have a small, albeit negligible, effect.

These attractive forces are strong enough to move mountains and exert a measurable effect on oceans, tugging the water one way when the sun and moon are close and the force is greater, tugging it another way as the force is relaxed or moves to another part of the sea. Since water is a fluid, if it rises at one point, it must lower elsewhere, causing a current to flow from one point to the other. If left undisturbed, any body of water will eventually settle into a stable configuration with a flat surface. The surface tension of the fluid acts to preserve that stability. It is only in the presence of some disturbing force that the surface begins to change shape.

WAVE TYPES

Wind is the usual disturbing force, producing wind-generated waves, the waves most commonly experienced. Oceanographers distinguish six major types of waves, typically categorized in terms of increasing period (Figure 1). These are capillary waves, which have periods of a fraction of a second; ultragravity waves, which have periods ranging from 0.1 to 1 second; gravity waves, 1 to 30 seconds; infragravity waves, 30 seconds to 5 minutes; long-period waves, 5 minutes to 24 hours; and transtidal waves, 24 hours or longer. In terms of the topic of this book, wind-generated gravity waves are the most likely source of extreme waves, but the next category, long-period waves, can also produce extremely dangerous and damaging waves.

Gravity waves—the name arises from the fact that wind piles water up, but gravity pulls it back down—are further subdivided into *seas* and *swell*. Seas—wind-driven waves—usually have shorter periods and are of greater significance to the mariner. Swell—a succession of waves that have moved beyond the immediate influence of the wind that caused them—have longer periods (10 to 30 seconds), but there are overlaps and no distinct dividing line between periods. Because the wavelength of swell is usually long compared to the length of a vessel, it generally does not concern the mariner in the open ocean. When the swell is large and its wavelength close to the length of a vessel, it is a different matter.

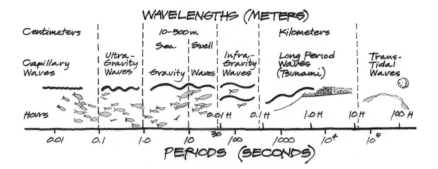

FIGURE 1 Types of ocean waves.

Other natural forces can disturb the equilibrium of the sea surface, wind being only one of them. These forces may be as small as the wake of a passing ship or as large as the great currents that circulate through all of the major oceans, the horrifying force of major storms, undersea earthquakes and landslides, and the tides themselves.

WAVE PROPERTIES

Some of these forces primarily affect the surface of the oceans, which is of great importance, of course, because it is where most human activity takes place at sea. Still others can move a body of water extending all the way from the bottom of the sea to the surface, and this difference is of vital importance when such disturbances approach shallow water. An important parameter for describing waves is the *wavelength*, L, which is simply the distance between two successive wave crests or peaks. The ideal wave can be visualized as a sine wave—that is, having the same shape that can be produced by tying a rope to a tree and shaking the free end up and down vigorously, causing a series of S-shaped waves to travel down the rope. Such motion is described by three properties: in addition to the wavelength, there is the *wave height*, H, the height of the crest above the trough, and the wave *period*, T, the time it takes to make one complete cycle. Sometimes the wave *amplitude*, A, is used; this is the distance from the centerline to the crest. A is equal to one-half H for the ideal wave. Whenever wave height is mentioned in this book, it will refer to H, the crest-to-trough distance. The period is the time in seconds for one complete cycle of a wave—that is, from one crest to the next.

The *wave speed*, c, is also of interest. It is equal to the wavelength divided by the period—that is, the distance traveled divided by the time it takes to travel. Alternatively, we can write:

$$\text{Wavelength} = \text{Period} \times \text{Wave Speed}.$$

This is a fundamental relationship that applies to many different kinds of waves, including light, radio, electrical, and sound. The main features of a wave are summarized in Figure 2.

In *deep* water, the speed of the wave is independent of depth and is determined only by the period. Thus, a wave with a period of 12 seconds and a wavelength of 750 feet will have a speed of 750/12 = 62.5 feet per second (37 knots). As this same wave moves into *shallow* water, its speed depends on the depth of the water. For example, if the water depth is 11 fathoms (66 feet), the wave speed is 27.2 knots, but when the depth is 5.5 fathoms (33 feet), the wave speed drops to 19.3 knots, and so on. As the wave slows, its height builds. Consequently, shallow water affects both the height and the speed of a wave—a most important thing to remember.

The energy carried by a given set of waves is proportional to the wave height squared. There are more exact equations, but we need not be concerned with them. It is sufficient to know that local newspapers run articles when winter storms bring waves 10 feet high to the beaches in Newport Beach, California, where I live. Yet a 100-foot-high extreme wave packs 100 times as much destructive potential as the 10-foot-

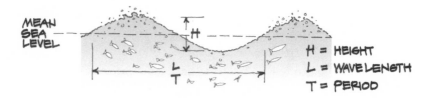

FIGURE 2 Properties of an ideal wave.

high waves that we would find very impressive while watching the surf crashing ashore.

Meteorological reports typically provide wave periods (T) and wave heights (H). The speed, period, and wavelength of an "ideal" wave are related mathematically. However, it should be understood that the periods and wave heights given in weather reports are "typical" values, meaning they represent the preponderance of waves of various sizes and periods that the mariner may encounter. Thus, the most likely height wave and the most likely period wave bear no mathematical relationship to each other.

The wavelength is related to how fast waves travel but has another useful purpose as a measure of when water is deep for a particular wave. As a rule of thumb, oceanographers say that water is deep if the depth is greater than one-half a wavelength. Since the average depth of the oceans is around 2.5 miles (13,115 feet, or 4,000 meters), for most waves in the open ocean the water is deep and the effect of the wave does not extend more than 300 to 600 feet below the surface. A submarine at this depth would not sense the turbulent seas of a major storm above it; here the sea remains calm. However, there are certain waves— earthquake-produced waves known as *tsunami*, for example—that have very long wavelengths; thus, even midocean is considered "shallow" for these waves. The passage of such long-wavelength waves roils the entire sea, from the surface to the depths, but this is generally of no major consequence until the wave enters shallow coastal water, where it slows down and piles up into towering and destructive walls of water capable of sweeping all before them.

Another important wave parameter is wave steepness. The steepness is a measure of how dangerous a wave can be. Thus, a very high wave with a short wavelength is said to be very steep. Wind or sea bottom conditions can increase steepness. In contrast, a high wave with a very long wavelength (such as an ocean swell) presents less of a threat to vessels since they can ride up and over the gradual slope. Theoretical studies of the idealized wave shown in Figure 2 indicate that there is a limit to the steepness of a wave traveling in deep water. Once the height of the wave reaches one-seventh of the wavelength, the wave is so steep that it will break.[2]

Back onboard *Dreams*, the view from Sixty Mile Bank revealed an endless expanse of blue water. We were too distant from land to view the shore; no passing boat interrupted our feeling of complete isolation on the fringe of the Pacific, the world's largest ocean. The experience of floating in the calm sea that day brought to mind the beginning of life itself.

OCEANS AND SEAS

The primordial forces that created the earth and left nearly three-fourths of its surface covered by water made life possible. Liquid water is one of the features that differentiate the earth from the lifeless planets in our solar system.

As large as the land surfaces may seem, they are dwarfed by the vast oceans and major seas of our planet, occupying 139 million square miles.[3] (See Table 1.)

The reach of the oceans is perhaps best visualized by realizing that the Pacific Ocean alone covers 32.4 percent of the earth's surface, more than all of the landmasses combined. (Here, where we calmly floated during our journey from Marina Del Rey to Baja, we were a microscopic speck on the largest segment of the earth.) The Atlantic Ocean

TABLE 1 The Earth's Oceans and Seas

Water body name	Area (million square miles)	Average depth (feet)
Pacific Ocean	64.0	13,215
Atlantic Ocean	31.8	12,880
Indian Ocean	25.3	13,002
Arctic Ocean	5.44	3,953
Mediterranean Sea	1.15	4,688
Caribbean Sea	1.05	8,685
South China Sea	0.90	5,410
Bering Sea	0.88	5,075
Okhotsk Sea	0.61	2,749
Others	7.9	—
Land	58 (29%)	—
TOTAL	197	

is the second largest of the great bodies of water, covering 16.2 percent of the earth, and the Indian Ocean is third, covering 14.4 percent.

What, exactly, is an ocean? Oceans are vast bodies of saltwater that separate continents. Far to the west of us, after passing innumerable Pacific islands and remote atolls, the Pacific washes up on the shores of the continents of Asia and Australia. Although we commonly interchange the words "ocean" and "sea," seas are in fact parts of oceans, but you must pass through some strait to reach them—the Strait of Gibraltar from the Atlantic Ocean to the Mediterranean Sea being one example. Or, farther south along the course we were pursuing, at the tip of Baja California lies the entrance to the Sea of Cortez.

For roughly 5,000 years, people have ventured into the vastness of the oceans and seas, seeking food, trade, a new place to live—or just yielding to the urge to explore the unknown. As long as 4,500 years ago, the ancient Egyptians built vessels that traversed the Mediterranean Sea to places as distant as Lebanon to bring back lumber and other trade goods. And, from the earliest written records, the dangers inherent in the sea—storms and giant waves that crushed frail vessels—are evident.

In *The Odyssey*, Homer (eighth century B.C.) describes how Odysseus, after offending Poseidon, god of the seas, witnesses his raft destroyed by a mighty wave:[4]

> With that he rammed the clouds together—both hands
> clutching his trident—churned the waves into chaos, whipping
> all the gales from every quarter, shrouding over in thunderheads
> the earth and sea at once—and night swept down from the sky—
> East and South winds clashed and the raging West and North,
> sprung from the heavens, roiled heaving breakers up—
>
> . . . At that a massive wave came crashing down on his head,
> a terrific onslaught spinning his craft round and round—
> he was thrown clear of the decks—

The story has a happy ending, however, because Ino, a sea nymph, takes pity on Odysseus and gives him her veil as a life jacket. He strips off his clothes and manages to swim to shore as his raft is lost. Unfortunately, the outcome for many other victims of ships struck by giant waves is not as happy; numerous ships have disappeared leaving no survivors, only pieces of floating debris to mark their demise.

A BRIEF HISTORY OF OCEAN EXPLORATION

Despite the dangers, humans continued to explore the oceans for a thousand years beyond the time of Homer, staying close to shore at first, then gradually venturing farther and farther offshore, seduced by the traditionally calm waters and idyllic weather of the Mediterranean. In 1000 B.C. the Phoenicians took control of the Mediterranean and made it a Phoenician lake. If we believe Herodotus, a Phoenician crew was the first to circumnavigate Africa, taking three years to pass from the Red Sea, down the east coast, around the Cape of Good Hope, and back through the Straits of Gibraltar.[5] From Scandinavia, Viking raiders reached England, France, and Spain, and traveled east to parts of Russia and coastal areas of the Baltic Sea. Piloting sturdy but light oceangoing vessels 60 to 80 feet long, they crossed the North Atlantic Ocean to Iceland and Greenland, and reached North America (Newfoundland) around A.D. 1000.

At this same time, but half a world away, China was turning out the world's best sailors and navigators. The Chinese developed paper, produced accurate nautical charts, used astronomy for navigation, and began to explore the South China Sea in the most reliable oceangoing vessels built up to that time.[6]

When Ming Dynasty Emperor Zhu Di came to power, he placed a man named Cheng Ho in charge of a shipbuilding program. At a shipyard in Nanking, Ho saw to the building of hundreds of vessels—some nearly 500 feet long—to create a Treasure Fleet that was to explore all of the known oceans and develop trade with foreign nations. Cheng Ho led seven expeditions between 1405 and 1433, traveling south to cross the Bay of Bengal, the Arabian Sea, and the Indian Ocean and, to the north, the East China Sea and the Sea of Japan. In the west, he reached the east coast of Africa, near the site of Mombassa, and in the northwest, he traversed the Red Sea as far as Mecca and into the Persian Gulf. He also visited Sumatra, Indonesia, Thailand, and Borneo. He chronicled rough seas and vessels lost on his voyages. In 1424, the emperor died and his successor closed the shipyard and idled the fleet. The Chinese withdrew into isolation.

Arab traders took the place of the Chinese and by A.D. 1400 were in control of the trade routes and principal ports in the Indian Ocean.

Their range extended from the Red Sea and Persian Gulf south to eastern Africa and the Spice Islands, but they dared not venture into the unknown of the South Atlantic, due to the fearsome waves of the Agulhas Current.

Meanwhile, first the Portuguese and then the Spanish were exploring southward, establishing bases along the west coast of Africa, and in 1488 Bartolomeu Diaz rounded the Cape of Good Hope. This set the stage for Vasco da Gama to round the Cape and finally reach India (1498), in an effort to reduce dependence on Arab traders for spices and other valued products from the Orient. Portuguese success led to Spanish concerns that the Portuguese would dominate the lucrative trade with east Asia, and eventually prepared the way for Columbus to prevail in his quest for royal approval of a voyage west—the "backdoor" route to the Spice Islands and the riches of the Orient.

DISCOVERERS OF NEW WORLDS—
COLUMBUS AND MAGELLAN

Along with his discovery of the New World was Christopher Columbus's great achievement of recognizing that there was something different about the winds that originated around the Canary Islands. Here were winds that were westerly and constant, as opposed to the variable winds encountered sailing down the coast. Without realizing it, Columbus had discovered the Northeast Trade Winds.[7]

Columbus's passage across the Atlantic was remarkable. His log shows days of steady sailing at speeds of 6, 8, and even 10 knots through calm seas. No major storms were encountered. Columbus knew that he could return to Spain by first sailing northeast to catch the westerlies—the winds that blew east across the Atlantic, and that is exactly what he did.

Fernão Magalhães (Magellan) was Portuguese and, like Columbus, sought the support of Spain when Portugal refused to support his grand plan to go east by sailing west. Eventually he secured the support needed, and on September 20, 1519, he sailed from Spain into the Atlantic with a fleet of five vessels, known thereafter as the Armada de Molucca. Six months later, battered by recurring storms and waves that had nearly sunk the fleet, and realizing he could not find the magical strait to the Spice Islands before winter fell on his fleet in its full fury, he resolved to find a safe place to spend the winter. This turned out to be a protected harbor near the tip of Argentina, known today as Puerto San Julian. Besides weather, Magellan was challenged by mutiny, lost one ship in a storm, and had another ship desert and return to Spain. The remaining three vessels eventually traversed the strait that bears his name and made it across the Pacific, reaching Guam on March 6, 1521.

A year and a half later—on September 6, 1522—a heavily damaged vessel was observed approaching southern Spain. It was the *Victoria*, the last of Magellan's ships, his remaining crew finally making it home to tell the story of his epoch voyage. The other two ships had been lost in storms; Magellan himself had been killed and buried in the Philippines, the victim of an ill-advised fight with natives. The 18 surviving crewmen on the *Victoria* could claim the first circumnavigation of the world. Not only that, as proof of Magellan's acumen, they unloaded a cargo of 381 sacks of cloves, sufficient to make the trip profitable despite the loss of three vessels.[8]

Today, the oceans continue to serve as a vital component of human existence, not only as means of transport and sources of food, but more importantly as essential elements of the earth's climate control system, which makes the planet habitable. Water fills in deep canyons and broad plains, except in those areas where the continents or islands extend above the surface. The deepest points of the oceans surpass the highest points on land. The Challenger Deep, in the Philippine Sea, lies an incredible 36,000 feet (nearly 7 miles) below the surface of the sea! By comparison, Mount Everest, the highest point on land, is only 29,055 feet high.

MODELING WAVES

With the discovery of new worlds and the growing importance of maritime trade, better understanding of wind, waves, and weather took on new importance. Certainly the earliest shipbuilders concerned themselves with staying afloat, constructing vessels that would not swamp easily, and developing these vessels by a process of trial and error. As sailing vessels became larger and more complex, shipbuilders had to consider the impact of waves, but more importantly, they had to consider the impact of wind. There was little reason other than scientific curiosity to try to understand the processes by which waves form and travel across the ocean. In the 1800s, some early work was done by George G. Stokes (1819-1903), an Irish-born mathematician and physicist. But it was not until the time of the Second World War that more serious efforts began. As the war turned the oceans into vast battlegrounds, understanding waves suddenly assumed new importance. Not only were battles fought in places and at times in which nature was as much an adversary as the enemy, it became necessary to land troops on islands and beaches in the Pacific and in Normandy. This was difficult enough in flimsy vessels in the face of enemy fire, but in rough seas it became exponentially hazardous.

The earliest attempts to model wave actions in the ocean used the approximation that waves are sinusoidal—the idealized wave discussed at the beginning of this chapter and portrayed in Figure 2. At this point it is important to state that *there are no such idealized waves in the sea.* The sea is always in a chaotic state, the actual surface motion being a composite of hundreds of waveforms that have been reflected and refracted by distant islands or landmasses, intersecting and diverging to form a complex surface on which no two waves are the same. The next effort to represent realistic wave patterns was to consider waves as the sum of many idealized waves. In this approach, a number of sinusoidal waves with different heights and wavelengths were combined. This method enabled oceanographers to reproduce complex wave shapes, but it still falls short of serving as a tool that will work under all conditions—deep and shallow water, small and large waves. Now we know that it is possible to speak of realistic waves only in terms of statistical averages. For this reason, wave heights are reported as the *significant wave height,* H_s, and meteorological reports typically give the *dominant period.* The significant wave height is defined as the average height of the largest one-third of the waves, for a representative group of waves, and the dominant period is the period occurring most frequently in a group of waves. For the purposes of this book, I must define one other term: a wave abnormally much higher than the significant wave, which I call an *extreme wave*, H_{ext}.

During the Second World War it became important to understand not only how waves form and move, but also the link between various sea conditions and waves so that wave heights and periods could be forecast. In the early stages of the conflict, the U.S. Navy commissioned Harald Ulrik Sverdrup, the director of the Scripps Institution of Oceanography, and Walter Heinrich Munk, a researcher at the Scripps Institution, to come up with an answer to this problem. Their initial effort considered ocean waves as periodic motion, represented mathematically as the sine waves described earlier in this chapter. A number of assumptions must be made for this approach to be valid. One is that the amplitude of the waves is very small and that waves are unchanged as they propagate. This method also assumes that all waves are identical at one point in time.

It does not take much observation to understand that the conditions discussed above are rarely if ever encountered in the ocean, as Blair Kinsman points out in his classic work *Wind Waves*.[9] Instead, there are all manner of waves—some small, some larger, occasionally one much larger—and not all travel in the same direction or even at the same speed.

Yet despite their randomness, ocean waves do exhibit certain regular or reoccurring patterns and are thus amenable to applying *Fourier analysis* techniques to model them. Jean Baptiste Joseph Fourier (1768-1830) was a French mathematician who discovered that any periodic function could be approximated by a series of sine waves. The opposite is also true; by applying a Fourier transform, a complex function can be broken down into its components. This mathematical process was invented by Fourier in the 1700s. He had accompanied Napoleon's army to Egypt as a science adviser, saw the French win many victories, and then was stranded in Egypt when the British navy sank the French fleet at Alexandria. This was a classic battle in which the British used wind and wave to successfully attack the French fleet! Fourier eventually made his way back to France with the tattered remnants of the French army, and was happy to once again be a mathematician and teacher. He was not forgotten by Napoleon, however, who drafted him to serve as prefect of the department of Isère. Fourier never sought this position, but given no choice in the matter, he did a commendable job. In his spare time at Grenoble, he developed the details of his method for Fourier expansions and transformations. (See Chapter 3 for more details.)

A similar process is used in earthquake studies. Seismometers measure how the ground moves as the waves generated by earthquakes pass through the earth. The resulting record, typically a squiggly line called a *time history*, shows how the earth moved at any point in time during an earthquake. If this record is digitized and analyzed using Fourier's methods, a different type of graph is produced. The wave is broken down into component waves, and the resulting graph shows amplitudes of the components plotted versus frequency. This is useful because it shows at what frequency most of the earthquake energy is concentrated. If a nearby building has natural frequencies of vibration near the dominant ones in an earthquake, it will shake harder and more likely be damaged than if its frequency were different.

Similar data can be gathered by recording instruments placed on buoys in the ocean or on offshore platforms. The time histories of waves can be converted to frequency spectra that show where most of the energy lies. The predominant frequencies can likewise be converted to a dominant or typical wave period. Like buildings, vessels have characteristic natural frequencies that depend on their size and method of construction. When these frequencies are close to wave frequencies, the vessel will roll or pitch more, sometimes to the point of capsizing.

The next refinement in wind-wave theory was to recognize that due to the randomness of waves, probabilistic methods of analysis were necessary to make more accurate forecasts of wave heights and frequency of occurrence.[10] Today, many improvements have been made in the mathematical models. Oceanographers are collecting wave data at sea to check the mathematical models and, in some cases, are using wave channels (analogous to wind tunnels) capable of creating artificial waves in the laboratory. Still, there is a lack of measurements on extreme waves. Some records have been obtained, but because giant waves occur infrequently and without warning, as well as the sheer difficulty of making measurements in huge seas when people's lives are at stake, the amount of actual data is small.

Fortunately, where we drifted near Sixty Mile Bank that day, big waves—especially extreme waves—were not a consideration. The hurricane season was over and we did not expect any bad weather. Calm seas have a certain peacefulness, but being becalmed soon makes a sailor anxious. Eager to be under way, we searched for signs of wind.

2

The Four Winds
and Waves

efreshed by our midocean swim, we ate lunch. Midocean is an exaggeration; even though the water was around a mile deep where we were and there was no land in sight, we were barely on the fringes of the Pacific Ocean. Gradually the breeze picked up and the sea surface became rippled with capillary waves. Small wavelets appeared. Soon the sails stopped luffing and began to fill. *Dreams* gathered forward momentum. The wind had arrived.

Wind and current are important ingredients in the formation of extreme waves. At certain places in the world's oceans, wind and current run amuck with regularity. Understanding their patterns is important for avoiding danger spots. No one knows this better than those who sail the Sydney–Hobart race.

BASTARDLY BASS STRAIGHT

"Thursday, April 19, 1770. Had fresh gales at SSW and cloudy squally weather with a large southerly sea. Saw land at a distance of 5 or 6 leagues. The southernmost point of land at latitude 37 degrees, 58 minutes south, longitude 149 degrees, 39 minutes east, I named Point Hicks, because it was Lieutenant Hicks who first discovered this land."[1]

With these words, Captain James Cook recorded his discovery of the east coast of Australia in his sea journal. In his first voyage of exploration, he had crossed the Pacific to Tahiti and then, under secret British Admiralty orders, sailed due south to determine if there was a great southern continent south of latitude 40 degrees. Reaching this latitude, he continued west, disproving the notion of a mysterious continent, and more importantly, accurately charting the north and south islands of New Zealand. From New Zealand he continued west across the Tasman Sea until he discovered the southeast tip of Australia.

Although he did not know it at the time, Cook was in Bass Strait, the 100-plus nautical mile wide channel between Australia and Tasmania. From this location west, no point of land intervenes until you track nearly two-thirds of the way around the world, back to the coast of Argentina. It is the notorious Southern Ocean, and in the vast distance between the Cape of Good Hope and Southern Australia, great storms build and pour their fury through the narrow and shallow waters of Bass Strait and into the Tasman Sea.

Cook caught the tag end of one such storm; it propelled HM Bark *Endeavor* to the northeast, past Bermagui, Batesman Bay, Jervis Bay, and finally to a bay he described as "tolerably well sheltered from all winds, into which I resolved to go with the ship." Due to the wealth of new and unusual plants discovered there, he named it Botany Bay. Later colonists would settle on Sydney as a more favorable location.

In 1945, following the end of World War II, a small group of sailors decided to race from Sydney south to Hobart, the largest city on the island of Tasmania, a distance of around 628 nautical miles. The race effectively follows Cook's course, but in the opposite direction, and then continues further south across Bass Strait and along the east coast of Tasmania into Hobart. Since then, every year on December 26 (Boxing Day), the race has been run.

Some sailors consider the Sydney-Hobart race to be jinxed—every seventh year, they claim, the weather turns really nasty, as occurred in 1970, 1977, and 1984. The trend came to a halt in 1991, disproving the myth, but ironically returned with a vengeance in 1998.

Peter Lewis, a veteran of six races or, as he puts it, "more accurately, five and one-fourth," raced in 1981, 1982, and even in 1984, when 104 out of 150 boats, including his, retired from the race due to extreme weather. Not discouraged, he raced again in 1985 and then, under the press of other obligations, did not race again until 1994 and then again in 1998.

On race day, Saturday, December 26, 1998, the weather forecasts were not good. A low was developing, and the initial forecast was for gale force winds, 34 to 47 knots. By race time, some forecasts were predicting 50 to 55 knots, and a storm warning was issued. What concerned the meteorologists was a fast-moving low-pressure area, heading east through Bass Strait. If it developed, the racing fleet would be pounded as it came past the south end of Australia and into the open ocean. It is storms out of the southwest that create a problem in the race. When this occurs, storm waves interact with the Australia Current coming down from the north, creating a condition in which waves pile up into huge seas while traversing the shallow waters of Bass Strait.

Sayonara, a boat owned by Larry Ellison of Oracle Corporation, was first out of the harbor, followed by *Brindabella*, owned by George Snow. Both were big, fast boats—fiercely competitive—and favored to place first overall. The balance of the fleet—113 boats—followed. Peter Lewis was on a 43-foot-long boat called *Esprit d' Corp*. Initially the trip south was good, the fastest boats doing 20 knots. Lewis had gotten a private weather forecast and knew that they faced 50- to 55-knot winds that night—"Tough but to be expected in a Hobart," was the way he put it to me.

His recollection of the seas that first night as the barometer plunged and the weather got worse and worse was: "A long cold night on the wind, a lot of banging and crashing into 50 knots and a building sea. At daylight on a cold, blowing Sunday morning, the crew of *Esprit d' Corp* noted that the mast had bent during the night—not a good sign, particularly with more bad weather coming. These weren't ordinary waves," Lewis said. "They had 4 or 5 feet of white water on top and no back to them. They'd leap up and the boat would rise and then drop like a stone for 20 feet. After much deliberation, and being very mindful of the weather, *Esprit d' Corp* turned west and headed for shelter at Bermagui. At this point, we'd traveled about 160 nautical miles, a little less than a third of the race. We made Bermagui around 3:30 Sunday afternoon, just as the first Maydays went out over the radio. The fleet and the storm had arrived simultaneously in Bass Strait, and all hell broke loose."

The faster boats were slightly ahead of the storm. Although *Sayonara* took a beating and suffered some damage, she finished the race Tuesday morning and took first place overall. *Brindabella* arrived a few hours later. Ellison is reported to have said after the race, "Never in a thousand years would I do this race again." Only 43 vessels made it to Hobart. Others were not so fortunate, especially those sailboats that were in the middle of the fleet—they caught the worst of the storm. As conditions worsened and Mayday calls started being heard on the radio, many vessels—*Esprit d' Corp* among them—made the wise decision to retire from the race.[2]

Plate 3 provides a sense of the sea conditions during the race. It shows *Bobsled* being blown sideways by winds estimated at 80 knots as she was entering Bass Strait. Breaking waves reached 80 to 100 feet, but the worst was yet to come. Boats ahead of *Bobsled* and behind her were capsized, broken up, and sunk by huge waves a little later in the day. The *Sword of Orion* was hit by a wave estimated to be 40 feet high at around 5:00 P.M. Sunday afternoon. The vessel was rolled completely and dismasted. Two crew members connected by safety lines were swept overboard. One managed to get back on board, but the helmsman's safety line parted and he drowned.

Stand Aside was hit by a huge wave that literally crushed the boat and then rolled it. The cabin roof was stove in, bulkheads failed, mast gone, hull leaking water, batteries submerged, engine inoperable. The crew launched two life rafts, but one failed to inflate and the wind broke the tether that held it to the boat. The crew—many of them injured—bailed frantically and threw everything overboard to lighten the boat. Photographer Richard Bennett took an extraordinary photograph (Plate 4) from a small plane flying 1,000 feet above *Stand Aside*. It shows *Stand Aside*, the remaining life raft trailing behind; a red smoke emergency flare has just been fired. Note the huge wave—80 feet high—that just broke to the right of the vessel. What started to be a routine photography assignment became a life-and-death rescue mission as Bennett and his pilot helped direct rescue aircraft to *Stand Aside* and other vessels in trouble. Miraculously, all 12 crew members from

Stand Aside were plucked from the towering seas by helicopter. As the last man was being hoisted into the rescue helicopter, a giant wave could be seen towering over the boat; moments later *Stand Aside* was swept to destruction and sank.

Winston Churchill was hit by a wave twice as high as the waves they had been experiencing; it was 80 feet high and heavily damaged the boat, but the crew managed to launch two life rafts before she sank from sight. In the stormy seas, several crew members were swept from the life rafts and lost; the others were eventually rescued.

Sailing slightly behind *Winston Churchill* was Professor Peter Joubert's boat *Kingurra*. The boat was hit by a large wave and rolled 140 degrees before righting itself. Joubert was seriously injured, with broken ribs, a collapsed lung, and ruptured spleen. Three crew members on deck with harnesses and safety lines were plunged into the ocean. Two were recovered, but as an attempt was made to haul the unconscious third man (John Campbell) aboard, he fell out of his safety harness and drifted out of reach. He was miraculously rescued around 40 minutes later by a Victoria police airwing helicopter. The helicopter was equipped with a radio altimeter. Professor Joubert later described to me how the helicopter hovered at 100 feet while crewman David Key was lowered into the water to rescue Campbell. As the helicopter hovered, the pilot, Senior Constable Darryl Jones, observed a wall of water approaching. He made an emergency climb to 150 feet to avoid having the helicopter swept into the sea. As the wave passed under them, the altimeter recorded 10 feet, meaning that the wave height was 140 feet.[3] This may well be the highest wave for which any type of measurement has been made.

In all, seven boats were abandoned; five were damaged, capsized, and sank. Six persons died, but altogether 55 were rescued, either lifted off their disabled boats or plucked from life rafts. The helicopter pilots and other rescuers did a truly heroic job of pulling people from seas in which the waves raged 70 to 80 feet high. At the time the race was running, an oil platform called Esso Kingfish B, located in Bass Strait, recorded significant wave heights in the range of 20 to 23 feet and an extreme height of 33 to 36 feet.[4] By the time these storm-blown waves reached the Tasman Sea, they had more than doubled in height.

At the end of my conversation with Peter Lewis, I asked if he planned to participate in the race again. He laughed. "Yeah," he said. "Stupidity is another name for it."[5]

REAP THE WHIRLWIND

The sun's energy penetrates the earth's atmosphere, heating the earth's surface and the oceans. Air receives nearly all of its heat from contact with the earth's surface. More of this heating takes place at the equator, where the sun's energy hits perpendicularly, and less at the poles, where it strikes obliquely.

As the air heats up, it expands, rising into the upper atmosphere in the equatorial regions and sinking as a dense, colder stream at the poles. This process creates high- and low-pressure areas on the earth's surface, causing winds to blow where air flows from high-pressure to low-pressure areas. The pattern of wind flow is complicated by the interference caused by landmasses and by the fact that the earth rotates counterclockwise when viewed from above the North Pole. In other words, a point on the earth's surface in the northern hemisphere moves easterly when the earth rotates, as evidenced by the fact that sunrise in Los Angeles occurs after that in New York or London. The earth's rotation gives rise to a force known as the Coriolis force that deflects fluids moving over the earth's surface. In the northern hemisphere, an airstream flowing west will be deflected to the north, away from the equator, while an airstream flowing east will be deflected to the south, or toward the equator. In the southern hemisphere the resultant directions are just the opposite. The deflection is zero at the equator and strongest at the poles. As a result of these several effects, air circulates in complex patterns over landmasses and the oceans, generally following circular paths that are predominately east and west. An example of a wind pattern that is well known to sailors who cross the North Pacific is the Pacific High. Winds blow in a great circular flow from Alaska down the coast of California and from Mexico west across the Pacific to Hawaii. The return path swings far north, sometimes to Alaska, and then down the Pacific coast. The center of this circular pattern is a movable spot, roughly in the middle of the Pacific between the mainland and Hawaii. There the atmospheric pressure is at its highest and little or no wind is to be found, as many a sailor has learned to his or her dismay.

A similar pattern exists in the South Pacific, off the coast of Chile, in the North and South Atlantic, and in the South Indian Ocean. As stated in Chapter 1, Columbus noted this pattern and gambled that the Northeast Trade Winds (blowing westerly) would carry him all the way to Japan and China. He was right about the wind, but underestimated the distance to Asia and did not know that the continents of North and South America blocked his course west.

The winds that sailors sense are the surface winds—air currents that move along the bottom of the earth's atmosphere. In addition, there are other circulating currents of air, some of which are vertical, moving down to or rising up from the surface. These are called *Hadley cells*, after George Hadley (1685-1768), the British meteorologist who discovered them in 1735. There are three such cells in the northern hemisphere and an equal number in the southern hemisphere. Warm air rises near the equator and descends at 30 degrees north and south latitudes. Warm air rises again around 60 degrees north and south latitude, to descend once again near the poles. The area near the equator where this occurs is known to meteorologists as the Intertropical Convergence Zone (but is known to sailors as the *doldrums* because of light and variable winds), and at 30 degrees it is called the northern subtropical divergence zone (known to sailors as the *horse latitudes*, similarly for lack of wind). The origin of the term "horse latitudes" is a reference to the difficulties faced by early voyagers. Vessels were often becalmed so long that the horses being brought to the New World became part of the sailor's diet.

High above the surface of the sea, the air flow again becomes horizontal and is characterized by the jet streams that blow from west to east and are familiar to all who have made cross-country airplane flights. The complex pattern of vertical and horizontal wind circulation is repeated in both the northern and the southern hemispheres. Weather patterns are similar, except for one difference that is important to those interested in extreme waves: the weather pattern of Antarctica.

Because of its massive ice cover—9,000 feet thick at the South

Pole—Antarctica is consistently 11 to 14 degrees Celsius (20 to 25 degrees Fahrenheit) colder than the Arctic. Because of this, the wind patterns described above are all pushed slightly to the north, meaning that warmer weather and thus warmer water are not centered around the equator but are found at some distance above the equator. Hence, there are no hurricanes in the South Atlantic and more hurricanes in the northern hemisphere than in the southern. However, strong westerly winds blow over the Southern Ocean all winter long.[6]

THE WINDS OF TRADE

William Dampier (1652-1715), a British explorer and buccaneer, was the first to describe the trade winds and the fact that they blow consistently from the northeast in the northern hemisphere and from the southeast in the southern hemisphere. Dampier made many other remarkable discoveries, recognizing that the equatorial currents flow in the direction of the trade winds and that tidal streams flowing near shore are not the same as ocean currents.[7]

In addition to the trade winds, which for centuries facilitated sail-driven trade and transport (hence the name), the global heat engine creates other winds, some beneficial but others detrimental to the navigator. Those that are detrimental include the kamsin wind, which originates in the Sahara, blowing southwest across Egypt from April to June; the mistral, a violent northwest wind that blows down into the Gulf of Lyons; the pampero, a dry northwest summer wind that blows from the Andes across the pampas to the sea; the sirocco, from the deserts of North Africa toward Italy; and the Santa Ana, warm air from the Mojave Desert blowing west to the Pacific Ocean to give just a few of the thousands of names for local winds.

The circulation of the winds on the surface of the ocean and in the upper atmosphere is quite complicated, so a complete description is beyond the scope of this book, but one further distinction that is important to us is the variation of wind patterns with the seasons. From winter to summer there are notable shifts in the locations where some of the highs and lows of atmospheric pressure occur. These shifts have an important effect on weather patterns and therefore on wave forma-

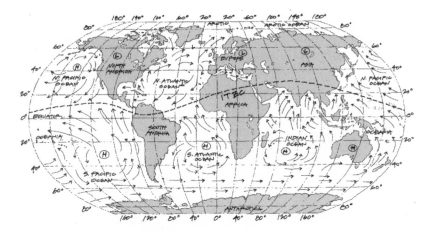

FIGURE 3 Ocean surface winds in August.[8]

tion. In keeping with the philosophy that "one picture is worth a thousand words," see Figure 3, in which typical high- and low-pressure areas are shown. In January, the lows over Siberia and Canada become
highs and the wind direction reverses in the North Pacific and Indian
Oceans. The dashed line shows the Intertropical Convergence Zone.
However, there is one final caveat: The figure shows the general trend
of the surface winds over the oceans, averaged over a long period of
time. Local disturbances and storms can at any time shift these patterns for a few days as local weather becomes extreme.

OCEAN CURRENTS

So, as in the earth's atmosphere there are great circulating currents, the
same is true in the oceans, where huge rivers of water, driven by similar
wind and density differences and modified by the Coriolis force, circulate. Currents initially flow in the direction of the wind but, under the
influence of the Coriolis effect, eventually deflect to the right in the
northern hemisphere and to the left in the southern hemisphere. This
deviation increases with depth, a phenomenon known as *Ekman transport,* after physicist V. W. Ekman who first described it. Another difference between currents and wind: When we speak of the direction of

currents, it is the direction *to* which the water is flowing. This is not true of winds. Traditionally—and this has always driven me crazy—when meteorologists speak of wind directions, as in "it was a terrible nor'easter," they are referring to the direction *from* which the wind blows. Throughout this book, I have endeavored to make it clear in which direction the wind is blowing, but beware the convention when you read weather reports.

Currents are very important to sailors, not just because of their role in causing huge waves on occasion—which is my purpose in discussing them here—but also because of their significance to navigation. Santa Barbara Island, part of the Channel Islands National Park, is located about 40 nautical miles southwest of Marina Del Rey, California. It is home to elephant seal, sea lion, and harbor seal rookeries.

I acquired a partnership in the sloop *Karess* with my friend Karl Bernstein before the advent of global positioning satellites. Our navigational aids were a compass and a radio direction finder. At that time, in 1987, Santa Barbara Island was a favorite destination for fishing and diving. On a good day, making 5 knots, it was an eight-hour trip over to the island, and we always left in the predawn hours. During the summer, the marine layer would often obscure the low-lying island until you were close to it; fog could obscure it entirely. (It is a small flyspeck of land, only 1 nautical mile wide.) Departing from Santa Monica Bay, *Karess* would be pushed sideways by the California Current, traveling south at a half to 1 knot. In navigational terms, we say that the current had a *set* (the direction toward which it is flowing) of 180 degrees and a *drift* (the speed of the current) of 0.5 to 1.0 knot. Unless adjustments were made, in just two hours of sailing, the current could carry the boat far enough to the south to completely miss the island—next stop, Hawaii, 2,200 miles away!

In the Pacific Ocean, there are two westerly flowing currents. The North Equatorial Current flows westward, driven by the Northeast Trade Winds from the west coast of North America across the Pacific to the Mariana Islands, where it veers to the northwest past the Philippines and Taiwan. The South Equatorial Current flows westward from South America, eventually curving southward into the Coral Sea and then south along the east coast of Australia. There is a similar pattern in the Atlantic Ocean. The North Equatorial Current begins near the Cape Verde Islands and flows west at an average speed of 0.7 knots.[9] The South Equatorial Current flows from the west coast of Africa to South America, commencing at a speed of around 0.6 knots but increasing to as much as 2.5 knots as it approaches Brazil. At the "hump" of Brazil (the state of Pernambuco), it divides—one portion going north; the rest, south.

Differences in water density (due to either temperature or salinity) will also create a current. Water flows from areas of lower density (where water depth is slightly greater) to areas of high density, where depth is less). As in the atmosphere, there are both surface currents and deep-ocean currents, again modified by the presence of landmasses that sometimes constrict or channel the flow. Analogous to the patterns of the wind in the atmosphere, these great currents flow generally in circular patterns in the world's oceans and are called *gyres*. Gyre circulation is of critical importance because it is a major mechanism for global heat transport.

In addition to the major currents described above, there are dozens of other named currents that profoundly affect navigation and coastal weather in specific areas. For the purposes of this book, some of the more important ones are shown in Figure 4. The numbered currents in particular should be noted, because they have special significance for the purposes of this book, as described later. To best understand how huge waves can arise, I believe it is important to be able to visualize not only the sea in motion but also the various forces that act on it to create extreme waves, particularly in those areas where of necessity oceangoing vessels must pass to make their way from one ocean to another, from one port to the next.

Oceanographers study current flow by using floating instruments with small position-indicating radio transmitters or by the simple expedient of dumping a lot of labeled bottles into the ocean. Recently, bad weather has expanded our knowledge of Pacific Ocean currents and provided data for comparison with computer models currently in use.

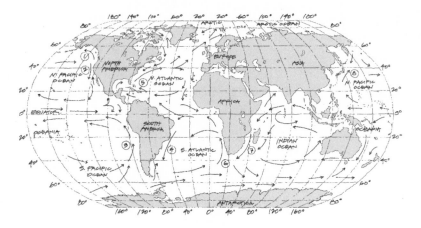

FIGURE 4 Major ocean currents. Note: (1) Alaska, (2) California, (3) Peru, (4) Brazil, (5) Gulf Stream, (6) Benguela, (7) Agulhas, (8) Kuroshio.

THOUSAND LEAGUE BOOTS AND RUBBER DUCKIES

On any day, hundreds of container ships cross the oceans. Major trade routes lie between Asia and the west coasts of North and South America in the Pacific and between Europe and North America and South America in the Atlantic. A modern container ship carries hundreds of 40-foot-long steel containers on deck. When battered by rough seas or struck by large waves, containers can be knocked into the sea, where they are floating hazards to navigation.[10]

An incident of this type occurred in May 1990, when the container ship *Hansa Carrier* lost 21 containers overboard during a storm in the middle of the Pacific, roughly midway between the Aleutian and Hawaiian Islands. These containers were full of Nike brand athletic shoes, and according to estimates, 80,000 shoes were dumped into the ocean.[11]

Remarkably, they floated and were carried in an easterly direction by the prevailing North Pacific Current. About six months later they began washing ashore on the coasts of Vancouver Island and Washington State, roughly 1,400 nautical miles from the point where they had entered the water. They were in surprisingly good condition after their long voyage, and enterprising beachcombers began collecting them.

The Internet and other means were used to advertise the availability of these shoes at bargain prices and also to trade with others in order to find a matching shoe in the same size.

Curtis Ebbesmeyer, an oceanographer with Evans-Hamilton Inc., learned that beachcombers were finding free shoes and immediately recognized that fate had created a superb North Pacific Current measurement experiment. Announcements appeared in the local press for people to report shoe findings and to include in these reports the date, location, and an identification number from the shoe. From the records of hundreds of individual "thousand league boots," researchers compiled a detailed profile of current speed and direction that showed reasonable agreement with the National Oceanic and Atmospheric Administration's computer model of Pacific Ocean currents.

A year or so later, some shoes reached the Big Island of Hawaii. At this point, they were truly "thousand league boots."[12] Ebbesmeyer predicted that if others remain intact in the ocean long enough, some will reach Asia and Japan.

Other container spills have added further confirmation of ocean current behavior. Ebbesmeyer reported a spill of 12 more cargo containers in January 1992—one of which contained thousands of small floating bathtub toys. Small yellow ducks, green frogs, blue turtles, and others successfully "swam" across the North Pacific and reached the beaches near Sitka, Alaska. Eventually some may be carried north and then east by the Alaska Current, eventually becoming part of the Arctic ice pack. In 1994, thousands of hockey gloves were lost at sea; they are now following the course of the bathtub toys. And in December 2002, a container ship ran into heavy weather off Cape Mendocino, California, and lost several containers overboard, releasing more Nike shoes into the Davidson Current. A month later, these shoes started coming ashore along the Washington coast after a journey of some 450 nautical miles.[13]

WIND WAVES

Knowing that the earth rotates around the sun, its surface largely covered with seawater in constant motion due to gravitational forces and temperature differences from the poles to the equator, it is not surprising that waves arise. When seas in constant motion come under the added influence of strong winds circulating above them, the stage is set for the creation of waves. Under the right circumstances, waves larger than you can imagine will occur, often with no prior warning, as Homer noted in the selection from the *Odyssey* quoted earlier, when ". . . East and South winds clashed and the raging West and North, / Sprung from the heavens, roiled heaving breakers up—."

It is at the interface of these four winds and water that waves are produced. Wind is not the only force that can produce waves, but it is the predominant one, and therefore it is important to understand how wind causes waves. As wind blows over the sea's surface, there is some friction between the moving mass of air and the water beneath it. The effect lessens the farther away the wind is from the interface; thus, winds at high altitudes do not have much effect on the water. Likewise, a violent storm will scarcely be felt beneath the surface of the sea.

Under the influence of the wind on still water, small water droplets are accelerated in the direction the wind is blowing. (You can observe this effect by blowing across a saucer filled with water.) In the sea, the wave direction may not be the same as the wind direction. Very light winds will produce small waves, a few millimeters high and a few centimeters long—known as capillary waves. These waves move at various angles to the direction of the wind, giving the sea's surface a characteristic wrinkled, diamond-shaped appearance. As the wind increases slightly, small riffles—sometimes called cat's paws by sailors—will be seen on the surface, traveling at an angle of about 30 degrees to the wind. Cat's paws are indicative of winds with speeds in the range of 4 to 6 knots. Some say that the sea's surface has a chicken wire appearance. If the wind dies, capillary waves fade away and the riffle smoothes out. Should the wind speed increase beyond a few nautical miles per hour, *gravity* waves are formed. If the wind continues to increase, they grow still larger. Now the waves move in the direction of the wind. As the waves become higher, the ocean surface becomes rougher, causing more wind turbulence and further increasing the wave height. This is also where *fetch*—the distance of unobstructed ocean over which the wind blows—becomes important. The longer the fetch, the higher the waves become. Long gravity waves are different from capillary waves in that if the wind dies, they keep traveling until they finally run into some obstacle, usually a shore.

WIND AND WAVE SPEED IN KNOTS

Nautical miles per hour is used universally to describe the speed of wind, waves, and boats (although meteorologists use meters per second). By long-standing tradition, nautical miles per hour is commonly abbreviated as *knots*. The word "knot" refers to the log line method sailors used to determine boat speed before the advent of instruments.[14]

Knowledge of a vessel's speed was essential for navigation; the speed multiplied by the hours of sailing gave the distance traveled from the last known position. If this information was combined with the vessel's direction (known from the compass heading) and corrected for any offset due to the set and drift of the current, the navigator could estimate the vessel's new position using a process called *dead reckoning*. To determine the vessel's speed, the navigator would periodically (say, once per hour under steady winds) throw a log line overboard and determine the vessel's speed. At best this was an approximation, since it determined the vessel's speed through the water but not over the ground. For example, if the vessel is making 6 knots into a current flowing 1 knot in the opposite direction, the true speed over ground is only 5 knots. After 24 hours of sailing, without correcting for the current, the navigator would erroneously believe his dead-reckoned position to be 24 nautical miles farther along his line of travel than it actually was. Thus, knowledge of currents was very important to the early explorers.

As the wind speed increases, or as the wind blows for a longer time, the sizes of waves increase. Wind acts on water in several ways. The first is by means of friction, as described above. The second is by a direct push. As waves build up, they form a vertical surface (the back of the wave) upon which the wind can act in much the same manner as a sail.

Only within the last 50 years or so has it been possible to develop mathematical theories that model the wind-wave interaction with any accuracy. For our purposes, it is sufficient to say that as wind first flows over the sea's surface, a "resonant" effect occurs that causes small wavelets to form. The resonant effect is not strong enough to cause waves to build. For this to occur, a second effect is required. As waves build, their very presence modifies the flow of air over the sea surface. Now, instead of being smooth, air is flowing over an uneven surface, and at the air-water interface, each fluid affects the other. To model this effect, analysts consider the sea surface "roughness," which can be characterized by a *drag coefficient*. The drag coefficient is a measure of the energy transferred from the wind to the water.

HOW WIND WAVES GROW

When all of the theoretical considerations are combined, it is found that wave height basically depends on the wind speed, the length of time the wind blows, and the unobstructed sea distance over which the wind blows.[15]

For a given wind speed, duration, and fetch, a maximum wave will be produced. Let us assume that we are in the open ocean and the wind blows steadily at 30 knots for 24 hours. Waves will build gradually until they reach a significant wave height of 19 feet and a maximum of 34 feet, as indicated in Table 2. Waves cannot continue to build in height indefinitely because at some point they start breaking and their energy is dissipated. When this balance is reached, the condition is referred to as a *fully developed sea*. If high winds blow over a narrow body of water, the waves will not be as large as the theoretical maximum and are said to be *fetch limited*. Likewise, if the duration of the wind is insufficient, the maximum conditions will not be achieved.

Plate 5 illustrates two wind-wave conditions. The top photograph shows waves resulting from Force 6 winds at 25 to 30 knots in the Gulf of Santa Catalina. The lower photograph shows waves during a Force 8 gale in the North Pacific.

Usually seas consist of many different waves, different periods and different wave heights, and waves traveling in different and similar directions. For this reason, marine weather reports give the significant wave height (H_s), defined in Chapter 1 as the average height of the

TABLE 2 Wave Height Versus Wind Speed, Duration, and Fetch

Average Wind Speed (knots)	Significant Wave Height (feet/meters)	Maximum Wave Height (feet/meters)	Significant Wave Speed (knots)	Significant Wave Period (seconds)	Minimum Fetch or Duration (nautical miles/ hours)
10	4/1.2	7.2/2.2	17	5.5	8.6/2.4
20	8/2.4	14.4/4.4	22.3	7.3	59/10
30	19/5.8	34.2/10.4	38.2	12.5	243/23
40	47/14.3	84.6/25.8	54.9	18.0	613/42
50	55/16.8	99/30.2	64.2	21.0	1,227/69

largest one-third of the waves, for a representative group of waves. In Southern California waters, a typical afternoon coastal wind forecast is "10- to 15-knot winds, swell 1 to 2 feet, wind waves 2 to 4 feet." It is possible to estimate the maximum wave heights for a given wind speed, fetch, and duration.[16]

Table 2 shows that a 50-knot wind can produce an extreme wave 99 feet high—if it blows for almost three days across 1,227 nautical miles of open ocean. These are the extreme values—winds blowing for less time or over a reduced fetch would not normally be expected to produce waves this high. Here, interpret "normally" to mean "in the absence of other factors." Extreme waves can result from interaction between moderate waves themselves, between waves and an opposing current, or when certain other conditions are satisfied.

THE BEAUFORT WIND SCALE

The earliest attempt to correlate sea conditions with wave height was done in the early 1800s by a British admiral, Sir Francis Beaufort (1774-1857). He produced a table in 1805 that was a masterpiece of clarity for ships in the days when shipboard instrumentation was uncommon. It was adopted by all the major seafaring nations. He divided the wind into 12 levels or "forces" as he called them, with Force 0 being "wind calm, sea like a mirror." Other examples include Force 5, "Fresh breeze; moderate waves, many white horses are formed," wave heights 6.5 to 8.2 feet, while Force 12 refers to hurricane conditions (winds greater than 64 knots) with seas 46 feet and higher.

When storms arise, mariners are advised by maritime authorities or national weather services through periodic weather broadcasts and by signals in harbors. There are four levels of warnings:

- Small craft advisory: winds up to 33 knots; one red pennant
- Gale: winds from 34 to 47 knots; two red pennants
- Storm: winds from 48 to 63 knots; square red flag with black square in center
- Hurricane: winds 64 knots and above; two square red flags.

During the day, flags are hoisted in harbors; at night a system of red and white lights is used to signal the warnings. However, in more remote areas there may be no local weather service and mariners must have the capability to access high-frequency radio weather faxes or Internet-based weather data via satellite.

LOWS AND HIGHS

Earlier I mentioned the Pacific High, a point at which there is little or no wind. Somewhere out in the middle of the Pacific Ocean—maybe at 145 degrees west, 40 degrees north—there is that spot where the barometric pressure (a measure of the weight of the earth's atmosphere bearing down on land and sea) hits 1,028 millibars, or maybe 1,030, or about 30.4 inches of mercury.[17] Virtually all of the weight of the atmosphere is concentrated in the first 19 miles (30 kilometers); above this altitude, the emptiness of space begins. The pressure exerted by the atmosphere was first measured in 1643 by Evangelista Torricelli (1608-1647), an Italian physicist. He constructed a mercury barometer, basically a long glass tube filled with mercury and closed at one end. The open end was placed in a bowl full of mercury, the closed end standing up vertically (Figure 5). The pressure of the air pressing on the surface of the bowl of mercury was sufficient to maintain the column of mercury in the closed tube to a height of 29.92 inches, or in modern terms a pressure of 1,013 millibars. This is considered standard atmospheric pressure at sea level and a temperature of 15 degrees Celsius.

Mercury barometers would be impractical on a vessel in constant

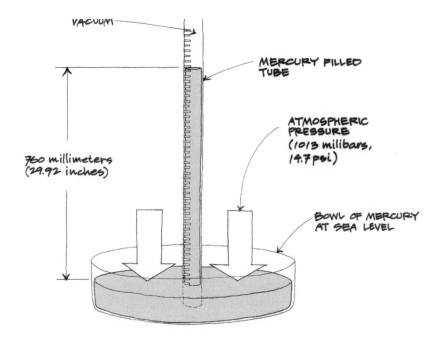

VACUUM

MERCURY FILLED
TUBE

ATMOSPHERIC
PRESSURE
(1013 milibars,
14.7 psi)

760 millimeters
(29.92 inches)

BOWL OF MERCURY
AT SEA LEVEL

FIGURE 5 How a barometer works.

motion, so for shipboard measurements an aneroid barometer is used. This is an evacuated metal box, normally round in shape, topped with a somewhat springy metallic diaphragm. The top is attached by a linking mechanism to a dial gauge. As atmospheric pressure increases or decreases, the evacuated box expands or contracts, and this movement is transferred to a dial calibrated in millibars and inches of mercury.

Since air moves from high pressure to low pressure, where the atmospheric pressure gradient is steep there will be winds, so mariners pay close attention to their barometers. However, most squall lines, almost all thunderstorms, and many other weather systems can hit suddenly with no warning on the barometer. Only in middle latitudes is the weather formed by large-scale traveling weather disturbances. In much of the world, pressure changes are minor to nonexistent when local storms approach. But when the barometer starts rising or falling, it is a sure sign of weather change. If the rate of fall is 1 to 2 millibars (0.03 to 0.06 inches of mercury) per hour, it can signal danger.

NOTABLE LOWS

In June 1996, a group of sailboats set out from New Zealand to Tonga. It was an informal regatta that included all types of boats, from monohulls to catamarans. The atmospheric pressure was 1,020 millibars. Without warning, after the boats were under way, a tropical depression began forming between Vanuatu and Fiji. The atmospheric pressure dropped suddenly from 1,001 to 986 millibars and the next day to 979 millibars. When the atmospheric pressure in a tropical depression drops at the rate of 1 millibar per hour for 24 hours, the New Zealand Meteorological Service refers to it as a "meteorological bomb." The gale from this depression moved southwest across the center of the fleet sailing for Tonga. Boats in the Force 12 storm's path were brutalized by high winds and violent seas. Five boats were lost, a number of rescues were carried out, and miraculously only three persons drowned. New Zealand Air Force planes involved in rescue attempts reported surface winds at 70 to 80 knots and wave heights up to 100 feet. The boats at the center of the storm reported Force 12 conditions. Boats were knocked down, dismasted, rolled over—some several times. Crews frantically reported their condition and location by shortwave radio to Auckland. A freighter in the area, a French navy ship, and New Zealand naval and air force personnel all came to the rescue.[18]

The crew of one disabled vessel was rescued by a large cargo ship. As the captain skillfully maneuvered the huge vessel close to the small sailboat, the crew waited for the exact moment when the mountainous seas crested, leapt from the sailboat to cargo nets hanging from the side of the larger ship, and then climbed as fast as they could to avoid being swept away by a giant wave. The jump had to be perfect—there would be no second chance in those roiling waters. With each passing wave, the two vessels ground together with a horrific noise as the sailboat drifted toward the stern of the cargo ship. When all of the sailors were safely aboard, they watched in horror as their beloved boat passed under the stern of the cargo ship. As the larger ship slammed down into the trough of a wave, it crushed the smaller boat, which promptly sank, carrying a family's belongings and memories of lengthy cruises to the bottom of the sea.

AROUND THE WORLD IN (LESS THAN) 180 DAYS

One area of the world where mountainous seas are routine is in the southern latitudes, 40 to 60 degrees south—known as the "roaring forties, furious fifties, and screaming sixties." Here, the winds can literally blow clear around the world, because there are no landmasses to interfere, so the fetch can be large. Near the southern tips of the continents the ocean is shallower, so the waves pile up and can be huge. Boats rounding these strategic points face another hazard: It is dangerous to go too far south because of the polar ice. Consequently any vessel making this journey runs a gauntlet of dangers. Yet from Magellan and the other earliest navigators to modern sailors in boats as small as 33 feet or less, men and women have challenged the capes. These races, in which sailors single-handedly sail around the world, are the world's most challenging athletic competition, with one's life on the line for endless days. Among the most dreaded and difficult portions of the route are the southern passages, where sailors face terrible weather, collisions with icebergs, and above all huge waves. You can imagine the strain of battling waves such as these, day after day, with virtually no sleep.

Ellen MacArthur, who came in second in the 2000-2001 Vendee Globe around-the-world solo race, experienced a knockdown during the race—one that snarled her mainsail near the top of the mast. She had to climb the mast while her 60-foot-long boat was surfing over 33-foot waves in a 40-knot wind. If the wind had increased before she could lower the sail, her boat could have suffered severe damage. But, having freed the sail, it took her more than an hour to climb back down, exhausted, freezing, at times swinging wildly and hitting the mast as the boat pitched and rolled.[19]

This experience served MacArthur well, though. In 2004 she took to the sea once again, this time in a 75-foot-long trimaran named *B&Q* (which she nicknamed *Mobi*). Her goal was to top the world record for a solo, round-the-world, nonstop voyage. Again, she experienced sail problems, and this time had to climb a 100-foot-tall mast to correct them. No problem; whatever it took, she did it—from hoisting a 300-pound sail, to repairing a generator, narrowly missing a collision with a whale, avoiding icebergs, and battling 50-foot-high waves. On February 7, 2005, at age 28, she completed the 27,000-nautical-mile trip in 71 days, 14 hours—a new world record.[20]

3

Over the Bounding Main

One summer night I anchored *Dreams* at Johnson's Lee, a sheltered cove on the west side of Santa Rosa Island, 120 nautical miles northwest of our home port at Newport Beach. Johnson's Lee was a submarine base during the Second World War, but the navy has since removed the last vestiges of the base. Santa Rosa Island is one of the northern Channel Islands and lies a little south and east of Point Arguello, where the California coast takes a sharp dogleg to the southeast. Here, the prevailing winds from the northwest funnel into Santa Barbara Channel between the islands and the mainland, often creating rough seas. This area is known to local mariners as Windy Alley.

After we finished dinner, the wind picked up noticeably, gusting on occasion to 35 knots. I watched small wavelets race across the cove, marking the passage of the wind. It was blowing out of the north, swooping over the island and down across the anchorage. I could hear the wind vibrating the rig and could feel the boat ride back as the force of the wind pulled up the slack in the chain that led to our 45-pound anchor and then move forward as the gusts died out. I thought briefly of setting a second anchor, but then decided we were hooked solidly. Bad decision!

Somewhere around 2 A.M. there was a huge gust—the captain of a neighboring boat thought that it registered 50 knots—and *Dreams'* anchor pulled free. We rode backward toward the open sea, dragging the anchor.

Words are inadequate to describe the chaotic scene that ensued: me jumping out of bed in my underwear, probing the darkness with a flashlight—the wind howling, the boat rocking, another crew member running forward to release the windlass and let out more chain, hoping the anchor would catch before we were blown completely out of the cove. Finally the anchor caught and I had a moment to clamber below for pants, shirt, and a jacket. After that, I maintained watch on deck until the sun rose and the wind died.

In the morning I surveyed the scene. The cove was surrounded by a dense growth of kelp. We had entered to one side where there was no kelp and anchored in a clearing broad enough for several boats. The wind had blown us backward several hundred yards through the kelp. Our path was clearly marked; it was as if a giant lawn mower had mowed a swath through the kelp.

A mariner's worst nightmare is to be blown onto a lee shore. We did not face that risk, since the wind was blowing offshore. Had the wind direction been reversed, I would have certainly set two anchors or, more likely, left the cove entirely for the open ocean. Still, it was enough—another dramatic and lasting lesson of how quickly the weather can change at sea. The wind shifts, waves build, and suddenly conditions are very different.

I spoke earlier of linear models used to analyze the "ideal wave." While these models lack the sophistication to accurately portray the random nature of real waves, they are useful in explaining certain phenomena we see in the ocean, and they are helpful in understanding how giant waves can suddenly emerge in seas that are average in size. A realistic situation is that one or more storms occur in distant parts of the ocean. Waves are created, and these disperse, the longer wavelengths running at faster speeds and leaving the shorter waves behind. The sudden appearance of these long, low swells that outrace the storm that caused them is sometimes a warning that a storm is approaching.

WAVE ADDITION

Suppose that at some point in the ocean the sea consists of only two types of waves with different wavelengths. When these two waves come together, they are said to either *destructively interfere* (cancel) or *constructively interfere* (superimpose, add together to create a larger wave).

Figure 6 illustrates these concepts. Destructive interference occurs when the wavelength and phase of the two waves are such that they effectively cancel out. On the contrary, if the two waves have similar wavelengths and are *in phase* (meaning that each successive crest of each wave lines up with the other), a wave with crests that are the sum of the crests of the two interacting waves will be produced. This is known as constructive interference.

Extend this reasoning to the real oceans, where there are countless unique waves coming from all points of the compass, originating from near and distant storms that happen every day over the world's oceans—you can see how it is possible for just the precise combination to occur occasionally, forming bigger waves. And at rare moments, even a wave so large that it will be called an *extreme wave* can form.

Now that we've ventured into the mathematics of waves, there are a few additional points that I want to make in the next page or two. After that, we'll leave mathematics behind us forever, cast overboard and bobbing in our wake, so to speak.

COMPOSITE WAVES

Since two waves can add together to form a larger wave, it is not difficult to imagine the opposite process—that of breaking down an irregularly shaped wave into a series of idealized sine waves, each with a different period, which added together exactly duplicate the original wave. Fourier, the French mathematician introduced in Chapter 1, developed this method, known as the Fourier transform. The only requirement is that the wave be periodic—that is, that it repeat itself at some interval.

For simple and regular mathematical functions, Fourier's method is straightforward to apply, but for a real ocean wave, application of the method before the advent of digital computers in the 1960s was te-

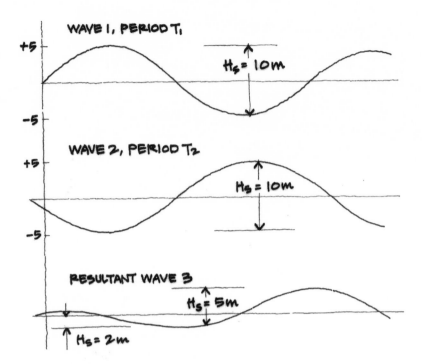

FIGURE 6A Destructive interference of waves.

dious. Computers made it possible to digitize the record of such a wave and then apply a new mathematical technique called the fast Fourier transform to break down the digitized wave into its constituent wave components.

For simplicity, the composite wave in this hypothetical example has only seven constituent waves (Figure 7). The transform process determines the frequency (reciprocal of the period) of each component as well as its amplitude or wave height. Since we know that the energy carried by the wave is proportional to its height squared, if we use the computer to calculate this quantity and plot it versus frequency, we can determine where most of the energy in the wave lies. This graph is called an energy spectrum of the wave and from it we can determine the significant period of the composite wave—that is, the wave period where most of the energy is concentrated. As noted in Chapter 2, this

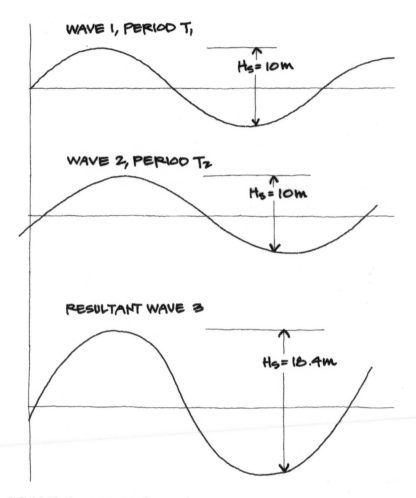

FIGURE 6B Constructive interference of waves.

method is similar to that used in the analysis of earthquakes and in calculating the response of buildings to a specific earthquake.

We've now seen how the seemingly endless variety of waves can be approximated by considering them to be the sum of a large number of uniform sine waves having different periods. In uncomplicated terms, this method is useful for forecasting sea conditions if some things—such as wind speed and duration—are known. However, as you might expect, this method is not useful for forecasting extreme waves.

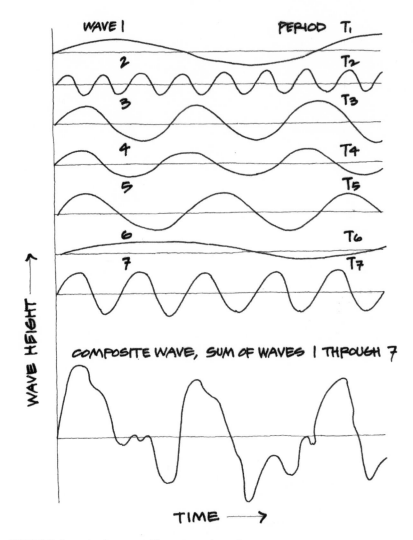

WAVE 1   PERIOD T_1

2   T_2

3   T_3

4   T_4

5   T_5

6   T_6

7   T_7

WAVE HEIGHT ⟶

COMPOSITE WAVE, SUM OF WAVES 1 THROUGH 7

TIME ⟶

FIGURE 7 A composite wave with seven components.

If one storm originates in close proximity to another, the combination can produce large waves, especially in the area of a strong current. This situation was tragically demonstrated in late October 1980, when a crew member on a Japanese tanker spotted a lifeboat floating in the sea. Inspection revealed that the damaged lifeboat had been torn from its supporting davits by a violent force. The lifeboat belonged to the *Derbyshire*, a bulk carrier—specifically, a type known as a combination carrier, which can transport ore, bulk goods such as grain, or oil. She was one of the largest ships of this type when launched in 1976. The *Derbyshire* was 957 feet long with a beam of 145 feet. When fully loaded, her freeboard, or distance from the water line to the top of the deck rail, was about 23 feet.

In September 1980, the *Derbyshire* was nearing the end of a voyage hauling 158,000 metric tons of iron ore concentrate from Canada to Japan. She crossed the Atlantic, stopped at Cape Town, and then proceeded for Japan under the guidance of a weather routing service. Approaching Japan in normal weather, the ship learned of an impending tropical depression and the weather routing service recommended a northerly course to avoid it. The *Derbyshire* increased speed to get ahead of the storm. On the following day, September 6, she reduced speed, apparently believing the danger had passed. Meanwhile, a second storm, Typhoon Orchid, arose close to the tropical depression, but the weather routing service apparently failed to advise the *Derbyshire* of this new threat.[1]

Typhoon Orchid, with winds of 85 knots and 60-foot-high seas, bore down on the vessel, now somewhere off Okinawa, and headed north in the Kuroshio Current. On September 9, the *Derbyshire* radioed that she was hove to due to a severe tropical storm. That was the last word that was ever heard from the *Derbyshire* or the 42 crew members and two spouses onboard. No Mayday or other distress signal was heard, suggesting that the foundering of the *Derbyshire* was sudden and cataclysmic.

When the *Derbyshire* failed to arrive in Kawasaki, a search was initiated. Search vessels and aircraft found a large oil slick around 300 nautical miles southeast of Japan but no other sign of the missing vessel or her crew. In 1987 the British government declared the vessel missing and presumed lost—"probably overcome by the forces of nature in Typhoon Orchid."

Shortly after the disappearance of the *Derbyshire*, several of her six sister ships were found to exhibit cracks in structural members and in the deck—in the aft portion of the ship, an area identified as "frame 65." In March 1982 the *Tyne Bridge* encountered severe weather in the

North Sea and, because several cracks were found in the deck, had to return to port. In November 1986, the *Kowloon Bridge* developed deck cracks in the same location while crossing the North Atlantic in a severe storm. She returned to port, slipped anchor, went aground, and broke apart near frame 65. These events suggested a flaw in the design of these bulk carriers and prompted family members of the *Derbyshire* crew to organize and fund an investigation to locate and survey the lost ship.

In May 1994, the *Derbyshire* was located in waters 13,700 feet (2.6 miles) deep, the bow section partially buried in an impact crater, the stern section some 2,000 feet away, and debris and iron ore scattered over the intervening area. With this new information, the British government launched a formal investigation into the loss of the ship, including two additional underwater surveys of the wreckage. A court hearing took place in April-May 2000.

The court found evidence indicating that heavy seas destroyed ventilators and air pipes on the foredeck, allowing water to enter the forward section of the vessel. The added weight forward caused the vessel to ride "bow down." Then with heavy seas crashing on hatch cover number 1, it collapsed, allowing more seawater to flood the vessel. Next, hatch cover number 2 likewise failed, and perhaps number 3, until the doomed vessel nose-dived out of control into the depths of the sea, finally impacting on the ocean floor at a speed of around 15 miles per hour.

The case of the *Derbyshire* illustrates how rapidly a storm can build and move. In the case in which the seas have already been built up by a previous storm, large waves can be produced. In the case of an extreme wave, a vessel could go bow down into the wave and never come out the back side. It was estimated that waves that hit the *Derbyshire* were at least 60 feet high. More significant, the *Derbyshire* was but another in a long list of bulk carriers that have been broken up or sunk by large waves. For additional details, see Chapter 10.

PROBABILITY OF LARGE WAVES

Wave heights are probabilistic—in other words, wave behavior can be analyzed statistically but cannot be predicted precisely. Weather forecasters are unable to say: "A 60-foot wave will occur at such and such a location at such and such a time," but they can say something like this: "There is likelihood that 8 percent of the waves that occur in a year at this location will be 60 feet or higher."

I stated earlier that weather forecasters report the significant wave height (H_S), which is defined as the *average* of the *highest one-third* measured wave heights (or the average of the highest one-third forecasted wave heights). Moderate waves in deep water can be modeled by a Rayleigh distribution (see Figure 8).

The vertical axis in Figure 8 is the *probability* that a particular wave height will occur, while the horizontal axis is the wave height. The figure shows the lowest 10 percent of the waves, the most probable wave height, labeled H_p, the average wave height H_A, the significant wave height H_S, and the highest 10 percent of the wave heights. [3]

While it may seem a little complicated, Figure 8 is one of the most important illustrations in this book. You can see that the significant wave height is equal to 1.6 times the average wave height. But still greater waves are possible, although they occur infrequently. The prob-

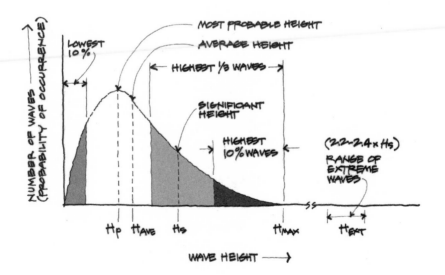

FIGURE 8 Rayleigh distribution of wave heights.[2]

ability of a wave *greater* than the significant wave is shown by the shaded area under the curve to the right of H_S. If the mathematics is carried out, this area is found to be 13.5 percent of the total, meaning there is a 13.5 percent chance that a wave higher than the significant wave will occur, or that roughly one in seven waves will be larger than the significant wave. Compare this to surfer's lore, as described in Chapter 6, or to the oft-stated mariner's view that every seventh wave in a set is larger.[4]

The maximum expected wave height is approximately 1.8 times the significant wave height. A mathematician would recognize that the Rayleigh distribution goes on forever. In other words, there is a vanishingly small probability of very, very large waves. If the wave height is much greater than this value, it falls outside the range we might reasonably expect. This is what has given rise to the term "rogue" or "freak" wave.

In a random sea, how many waves would it take before you experience the maximum wave? In 20 waves there is about a 5 percent chance of reaching a maximum wave, and in 200 waves a 5 percent chance of reaching a height twice as great as the significant wave height. An extreme wave—one greater than 2.2 to 2.4 times the significant wave—has a 5 percent chance of occurring in 1,000 to 4,000 waves if a constant sea is assumed.[5]

Under typical conditions, a vessel encounters 5 waves per minute, 300 waves per hour (waves with a 12-second period). Thus, in traveling 3.3 to 13.3 hours in such a sea, a vessel would have a 5 percent chance of experiencing an extreme wave. However, during this time, at 10 knots the vessel would have moved 33 to 133 nautical miles, hopefully to an area where the significant wave was smaller and the consequences of encountering an extreme wave less severe.

For larger waves, with variable wind conditions such as those associated with hurricanes, or for waves in shallow water, more complex methods are required and the analysis must be handled differently.[6] Waves in shallow waters and very large waves are nonlinear, meaning they have crests that are several times higher than the distance from the centerline to the bottom of the trough, and they are steeper than other waves.

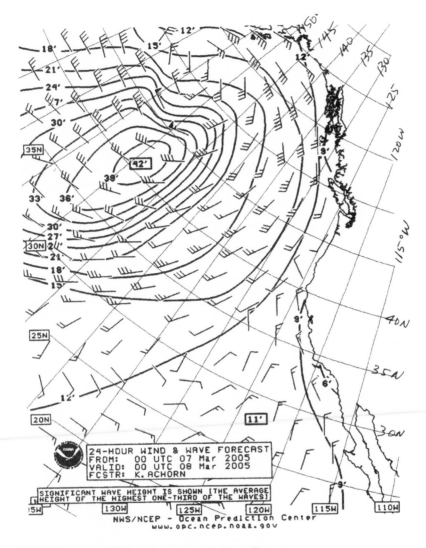

FIGURE 9 Typical NOAA marine weather wind-wave forecast.

Figure 9 is an actual U.S. Department of Commerce National Oceanic and Atmospheric Administration (NOAA) marine weather forecast for the North Pacific on March 7, 2005. Note the storm centered at
latitude 30 degrees north, longitude 150 degrees west, northeast of the
Hawaiian Islands. The forecast shows 40-knot winds and waves with a

42-foot significant height at the center of the storm. These waves had a period of 12 seconds. Looks like a spot to avoid, if you have the option. (See Chapter 5 for more about this particular storm.)

A new tool for measuring wave heights is the satellite. As a satellite orbits the earth and passes over the oceans, it collects a huge amount of data. The measurements are averaged, since it is not practical to calculate individual wave heights. Thus, these data provide reliable estimates of the significant wave height on a global basis, but not the extreme wave height. In Chapter 8, I describe more recent research into satellite observations that have been designed to detect extreme waves and the astonishing results that have been obtained.

Satellites reveal two major bands circling the world's oceans where the greatest values of significant wave height occur These are areas where the significant wave height is routinely almost 20 feet (6 meters). As you might expect, there is a distinct trend—there are more big waves in the winter. Figure 10 shows the situation in February—winter in the northern hemisphere. The darkest area corresponds to a significant wave height of 6 meters, the lighter shades, 4- and 5-meter heights. Note that they encompasses a broad area of the North Atlantic, from England to Newfoundland, and in the North Pacific, from the Gulf of Alaska to Japan, centered around latitude 50 degrees north. Meanwhile,

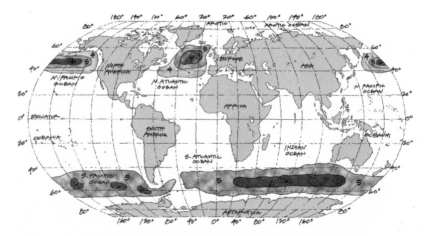

FIGURE 10 Global significant wave heights, H_s, in February (in meters).

at this time—summer in the southern hemisphere—the shaded area is centered along latitude 50 degrees south.

In August—winter in the southern hemisphere—the shaded band will have broadened into a swath of rough seas extending clear around the world, while in the northern hemisphere the shaded areas are gone. In other words, conditions in the Southern Ocean remain severe most of the year. Also, note the absence of large waves in the tropical zones near the equator. This area is subject to a different condition, that of tropical cyclones. These satellite observations are not programmed to detect such transient events as hurricanes. (Chapter 4 discusses wave heights generated by storms and hurricanes.)

If the world's major oceans are sliced along a longitudinal meridian in their center, the variation of significant wave height with latitude can be examined. If this is done, H_S increases from around 13 feet at 60 degrees north, to 20 feet at 50 degrees north, and then declines to 6 to 10 feet at the equator, again increasing to around 20 feet at 50 degrees south latitude. No wonder sailors call 50 degrees south "the furious fifties."

In the Southern Ocean, the wind is generally from the west to the east, the strongest winds occurring between South Africa and Australia, weakening as you move farther west. The data also show that the significant wave height is generally higher on the west coasts of Africa, Australia, and South America than on their east coasts. One exception to the general trend is the region at 20 degrees north in the Arabian Sea. Here, strong southwest winds that arise during the summer monsoons (June to September) create waves with significant heights in excess of 16 feet.[7]

WAVE MOVEMENT

How do waves propagate? Surprisingly, the water in a wave produced in Hawaii does not reach California; it is the wave motion that is propagated. If you place a cork in the path of a wave, it moves forward with the crest of the wave, but then moves *backward* in the trough until it rises and moves forward with the next crest. So waves actually move water up and down in circular orbits. A floating object will be seen to move up and down but will advance forward very slowly. Each circular orbit advances the wave only a short distance. This form of wave motion, which occurs in deep water, is called an *inverted cycloid*. To visualize it, imagine a point on the rim of a bicycle wheel—for example, the valve stem. Now picture the motion of the valve stem as the wheel rolls along a horizontal surface. From the top position it can be seen to roll forward initially, then when in the bottom position, to move backward relative to the bicycle's forward movement. In shallow water, the orbits of an individual particle of water change, becoming elliptical in shape. This results from the fact that as the water becomes very shallow the vertical component of motion decreases.

The same process occurs when a boat moves in the same direction as the predominant waves. The crest of the wave pushes the vessel forward, in effect adding the speed of the wave to the speed of the vessel. (Of course, due to the inertia of the vessel, it does not immediately reach the speed of the wave, which is generally moving much faster than the vessel.) As this crest passes under the hull, the boat enters the trough of the wave, where the direction of water flow is now *opposite* to the direction of the vessel, and the boat is slowed somewhat until accelerated again by the next crest. However, since the waves are propagating in the forward direction, the net effect is to increase the speed of the vessel over what it would be in calm water. Conversely, if the boat was moving against the direction of the waves, its speed would be reduced.

A wave rises up and then, as it collapses, displaces water, creating a new wave. Think of the ocean as a long tube completely filled with marbles, each marble representing a wave. Now jam another marble in one end of the tube, and what happens? A marble pops out of the other end—but each marble only moves a distance equal to its diameter. Likewise, one wave initiates another, which causes another, and the disturbance propagates at a speed we call the wave group or wave system speed.

Waves travel long distances, gradually dissipating energy as they move and also gaining energy from new sources of wind. As waves progress away from a storm or other initiating event, they gradually separate, the faster waves (those with longer wavelengths and periods) traveling farther in a given period of time. At long distances, waves will appear in groups, called sets—the long wavelengths appearing first, followed by the shorter wavelengths. This process is known as *dispersion*. As waves propagate they also *attenuate*, which is to say that they gradually lose energy, the short wavelengths attenuating more rapidly.

A wave system actually travels at half the speed of an individual wave. If you look carefully at waves approaching a vessel, you will see the oncoming wave crest and then disappear, transferring its energy to a wave that forms *behind* it; the speed of the entire wave system is only half that of an individual wave.[8]

Consider the North Pacific storm of March 7, 2005, described earlier. The wave height in the vicinity of Southern California's Channel Islands on March 7 was 6 feet. I checked the National Oceanic and Atmospheric Administration marine weather report for the same area on March 10. By that date the storm had moved northeast into the Gulf of Alaska and was located south of Valdez about 1,700 nautical miles distant from the Channel Islands, where the waves were now 12 feet. Since 72 hours had elapsed, these waves had traveled at around 24 knots to reach the coast of California.

Actual seas consist of many separate wave trains arising from different storms at varying distances, modified by refraction or reflection from nearby landmasses and by wave interference or superposition as described earlier. This is why the sea at any given time has a noticeably irregular surface, although superimposed on it are seemingly semi-regular patterns of the dominant waves emerging from dispersion. When wave systems cross at angles to each other, irregular peaks can be produced, or you might see the effect of two waves running into each other.

When the water becomes shallower—a process called *shoaling*—the wavelength and wave speed decrease and the wave height increases. In shallow water, defined as water where the depth is very much less than the wavelength, the wave speed and wavelength depend on the depth rather than on the wave period. In the nearshore area, as the wave slows, not only do the crests become higher, they also come closer together.[9]

BREAKING WAVES

As waves slow and become higher and steeper, their crests—less constrained by the ocean's bottom—move faster than their troughs. The result: the top of the wave topples forward, like a bucket full of water that has been tipped over. As any surfer knows, this process is called *breaking.*

Breaking waves are *plunging waves*—the type seen frequently in surfing films—the top curling over and plunging into the valley, forming a horizontal "tube" perpendicular to the wave direction; or *spilling waves,* wherein the wave shape is concave on each side, but rather than curling over, the crest slides down the face and disintegrates into a welter of foam. If waves do not break completely, they can reform and break again.

The shore is not the only place waves break. Strong winds can cause them to break in the open ocean. When the wind's speed is 30 knots, it blows spray from the tops of the waves. At Beaufort Force 9 wind conditions (41 to 47 knots), wave crests begin to topple over. Waves breaking in the open ocean are a definite hazard to vessels. If a 3.3-foot (1-meter) high column of water lands on a ship's deck, it exerts a static force of 1 ton per square meter (204 pounds per square foot); the dynamic load is even greater and damage can result. Another category is a *surging* breaker, one that rises to a crest but overruns the beach without plunging or spilling.[10]

The height of breaking waves is influenced by shore conditions, the direction of the incident wave, and the wavelength of the incoming wave. Longer waves will break in deeper water. On steeply sloping beaches, waves tend to be of the plunging type.

On a given day, walking between the Balboa Pier and the Newport Pier (both in Newport Beach, California), it is possible to see each type of breaker. There are sections of the beach where the bottom rises sharply, waves form plunging breakers, and there is a strong backwash of water returning to the sea, sometimes accompanied by riptides. Farther down the beach toward the Newport Pier, the bottom slopes more gradually and spilling breakers occur.

In Chapter 1, I stated that theoretical calculations predict that waves will break once their height reaches one-seventh of their wavelength. This suggests that to produce an extreme breaking wave with a height of 98 feet, the wavelength would have to be 686 feet or greater. An idealized (hypothetical) wave with this wavelength would have a theoretical speed of 35 knots and a period of 12 seconds. Such a wave would easily overtake most vessels in the ocean. There would be no outrunning it!

Another important wave parameter is the *steepness*, defined earlier as the height divided by the wavelength:

$$\text{Steepness} = \text{Height/Wavelength.}$$

Thus sailors are not concerned about high waves with long wavelengths because their vessel will ride up the crest and slide down the back side, with no perceptible discomfort if the wave is sufficiently long. It is the short-wavelength waves (or short-period waves) that create problems. In this case, a vessel climbing the crest of a wave will ride up and over, burying its bow in the trough of the next wave, taking on water and subjecting the vessel to severe impacts. Under extreme conditions, a vessel can *pitchpole*, or flip end over end in these conditions, often with fatal consequences (Figure 11).

While the speed of propagation of wind waves in deep water is determined by the wavelength, a different condition applies to tsunami waves. The initiating event for a tsunami is an underwater earthquake, landslide, or volcanic eruption that results in sudden displacement of a massive volume of water. Rather than the surface layer of water moving as in the case of wind waves, the entire volume of water from the bottom to the surface is accelerated by the initiating event.

Tsunami have extraordinarily long wavelengths (hundreds of kilometers) and periods of up to an hour. The speed is determined only by the depth of the ocean, and tsunami can cross vast expanses of ocean at speeds of 380 to 430 knots (437 to 495 miles per hour), or as fast as a jet plane. In the open ocean, the wave height may be 3 feet or less.

However, the volume of water that is put in motion is enormous. In a manner analogous to wind waves, when a tsunami approaches shoaling waters the wave slows, but because of its high speed, slowing has a much greater effect and the height of the wave can become 50

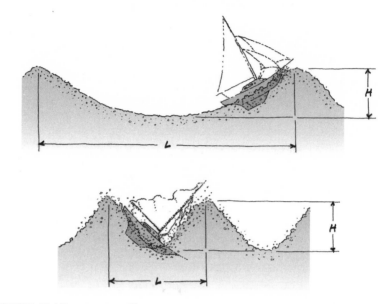

FIGURE 11 Wave steepness.[11]

feet, 60 feet, or more, causing extensive damage and loss of life. Usually there is more than one crest; the first crest may not be the largest. The arrival of the first wave may be signaled by a large and unusual back-wash that drains beaches and harbors, uncovering normally inundated surfaces, but this may occur after the first wave hits. In either case, spectators coming to view this unusual scene do so at considerable peril. (Tsunami are discussed further in Chapter 6.)

THE PULL OF THE MOON: TIDES

Imagine a wave that stretches halfway around the world, has a period of 12 hours and 25 minutes, and is moving at hundreds of miles per hour in the open sea.[12] As surprising as it might seem, you've experienced such a wave if you have spent any time on or near the ocean. The crests of this wave are known as high tides; its troughs, as low tides. Tide waves are not to be confused with tsunami although you will hear tsunami mistakenly referred to as "tidal waves"; however, they have nothing to do with the tides. Tides do not usually cause extreme waves, but there are two situations in which tidal forces can cause large waves. These are known as *seiches* and *tidal bores*.

Seiching can occur when any external force disturbs an enclosed body of water. Waves move back and forth from one end to the other. The period of the waves depends on the size (length and depth) of the body of water. Standing waves can also occur. Seiches can be caused in bays and harbors by tidal currents, by the arrival of a distant swell with just the right period, or by storms or a tsunami. Sometimes these will oscillate for days. Seiches are generally not a problem and are detectable only by means of careful measurements. However, in the case of harbor design, engineers determine the predominant wave periods in the area and ensure that the harbor dimensions do not create a condition in which large seiches can occur, since this could cause excessive movement of floating docks and straining of vessel mooring lines. On a large scale, seiches have proved dangerous, damaging boats at dock and occasionally killing people fishing near shores or on breakwaters. Such an incident occurred on June 26, 1954, when a 10-foot-high wave suddenly rolled in from Lake Michigan and swept eight fishermen off a breakwater, drowning them.[13]

TIDAL BORES

Tidal currents in the open ocean are weak—say, 0.1 to 0.2 knots—but near the coast or in bays and river mouths they can reach 5 knots or more. In shallow rivers, when currents exceed a critical speed, *tidal bores* appear.[14] This is one tidal phenomenon that can create unusual if not large waves—a wave that moves forward as a wall of water.

If large tides occur in narrow harbors or river mouths, the rapidly changing height of the water is compressed into a narrow channel, the current flowing much faster than the current of the tide wave in the open ocean. This can cause a fast-moving wave to sweep up the channel, sometimes creating hazardous conditions for boats entering or leaving the area. The wave can be a breaking wave or just an abrupt wave front; it is called a *tidal bore*. Usually tidal bores are around 3.3 feet in height, but they can be as high as 26 feet—for example, on the Qiantang River in China. Plate 6 is a photograph of the Qiantang bore taken in late August 1986 from a location near the city of Hangzhou. The bore forms below the city and then travels at a rate of 10 to 20 knots upstream for about 25 miles. In September and October, visitors arrive from all over China to watch the spectacle. A number have ventured too close and have been swept to their deaths. Local boatmen know to get their vessels out of the way!

WAVE BUILDUP NEAR SHORE

We had an unusual amount of rain in Newport Beach during the winter of 2005, and that plus a lot of travel kept me away from *Dreams* and offshore waters for most of the winter and early spring. Finally I saw an opening, and early in May, Nancy and I sailed to Emerald Bay, a small cove on the north side of Catalina Island near the west end. We planned to spend a few days before the Mother's Day weekend, when once again other commitments would prevail. Emerald Bay is a favorite spot of mine for a short trip because it is relatively isolated. The nearest neighbor is a youth camp on shore at the far west end of the anchorage. It is a pleasant spot for diving or hiking to Stony Point and an isolated beach called Parson's Landing. The weather was forecast to be mild but with a probability of showers, which was undoubtedly why we found ourselves the only boat in the anchorage. The cove has an entrance channel (fairway) about 1,500 feet long and moorings or room to anchor for about 100 small boats. Near the opening is a prominent rock—Indian Rock, about 1,000 feet from shore—marking the east end of a largely submerged reef that runs along the north side of the cove, giving some protection from swells coming in from the north. Just outside the reef the water depth drops quickly to 13 fathoms; and half a mile beyond that, it drops to 60 fathoms; and then another 2.5 miles out, it descends to depths of around 490 fathoms in the San Pedro Channel.

We arrived in the early afternoon, got the boat squared away on a mooring, launched the dinghy, and relaxed—me down below writing, Nancy on deck with her watercolors, sketching the island, now a brilliant green after all the rain. The moorings are in long rows; *Dreams* was on the second row, in about 3 fathoms of water, and about 160 feet from a rocky cliff that divides the beach into two sections. There was no wind and only a gentle swell from the northwest when we went to sleep. At 3:00 A.M. I awoke as the motion of the boat shifted. The dinghy tied alongside so as not to bang all night was in fact banging and the boat was rearing up and down like an impatient horse.

I went on deck to check the mooring lines and investigate. A breeze had come up—blowing over the island from the south—and this accounted for part of the noise. The main culprit, however, was a new swell coming from the north directly into the harbor. It was not terribly large—say, around a foot high, but the wavelength was about 33 feet—the same as the length of *Dreams*. As each wave came in, the boat would rise and then drop down into the face of the next incoming wave, creating the bucking motion that had awakened me. It was also just past low tide, and I knew the tide was rising. After making certain that all lines were secure and repositioning the dinghy so that it would make less noise, I went back to bed, making a mental note to wake up again if the swell increased, because it might become advisable to move out of the harbor.

I was remembering the story of the wreck of the *Memphis.*

In Chapter 2, I mentioned the buildup of waves approaching shallow shores and bays. The height that waves can attain under these conditions and the tremendous energy they carry was graphically and unforgettably illustrated in the wreck of the USS *Memphis,* a 500-foot-long armored cruiser.

On August 29, 1916, the *Memphis* was anchored in Santo Domingo Harbor in what is now the Dominican Republic. She was anchored in 7.5 fathoms (45 feet) of water, about a half-mile from a rocky beach. The harbor opens directly to the south and thus is vulnerable to storms. August is hurricane season in the Caribbean, so the *Memphis* maintained 2 of its 16 boilers in operation in case it had to leave the harbor in a hurry. A previous storm warning had shown that steam could be raised in 40 minutes when two boilers were left in operation.[15]

Shortly after noon on August 29, one of the ship's officers noted that the cruiser was rolling more than usual—about plus or minus 10 degrees. Captain Edward L. Beach, Sr., took a look at the horizon and could see a heavy swell starting to set in. The rocky shore now showed breaking waves. No storm warnings had been received, but the captain was so concerned that he issued the order to fire the boilers, rig the vessel for heavy weather, and be prepared to leave the anchorage. A second anchor was readied as a backup measure. A smaller naval vessel, the gunboat USS *Castine*, also made preparations to leave the harbor. The engine room informed the captain that sufficient steam would be available for the vessel to get under way at 4:35 P.M.

The timing of the following events is important. By 3:30 P.M. it was evident to the captain and other officers of the *Memphis* that conditions in the harbor were deteriorating. When they looked out to sea at around 3:45 P.M. they could see a huge wave of yellow water approaching them. The wave spanned the entire horizon and was estimated to be 75 feet high. By 4:00 P.M. the *Memphis* was rolling so violently that seawater was coming in through the ventilators 50 feet above the waterline. By 4:25 P.M., water was coming down the stacks 70 feet above the waterline—and was snuffing out the fires in the boilers.

Not only was the *Memphis* being battered by the waves, her only means of escape was being quenched by them as well. While all of this was transpiring, the *Castine* miraculously got under way and managed to get out of the harbor, pummeled and damaged by the large waves, but somehow avoiding capsizing.

By 4:00 P.M., the bridge had noted that the huge "ocher-colored wave" was nearer and had grown in size to 100 feet or more. Now the *Memphis* was rolling 45 degrees, causing the gun ports to go underwater and letting torrents of seawater into the ship. At 4:40 P.M. the ship started striking the bottom of the harbor. Finally the engines began to respond, but it was too little, too late. Waves were rolling the vessel and battering her on the rocky bottom, damaging the propeller shafts, and soon the engines lost steam pressure. Around the same time, the *Memphis* rolled over into a deep trough and then was hit by a succession of three large waves—the last one the largest. Water washed crew members overboard and flowed over all but the highest points on the ship. By 5:00 P.M. she had been carried near the cliffs and was resting on the rocky bottom. The photograph in Plate 7, taken by a bystander on the afternoon of August 29, 1916, shows waves pouring over the *Memphis*. In an hour and a half, she had been battered into a total wreck.

In the definitive account of the wreck, written by Captain Beach's son, the loss of the *Memphis* is attributed to a Caribbean tsunami. This did not seem plausible to me, given the time sequence summarized above. The succession of waves that hit the ship seemed to have periods more typical of storm waves. Also, the elapsed time from the sighting of the large wave and its arrival—something between half an hour and one hour—did not seem to indicate a tsunami. If you could see a 75-foot-high wave at a distance of 18 nautical miles, a tsunami traveling at 400 knots would have hit the ship in a little less than three minutes, and if it had slowed to 200 knots, in about five minutes.[16] On the other hand, if it was a storm wave traveling the same distance, the speed of 18 to 36 knots is about what you would expect.

I could find no record of a Caribbean tsunami on August 29, 1916, so I sought the help of Professor George Pararas-Carayannis. Pararas-Carayannis is one of the world's foremost authorities on tsunami, having served as the director of the International Tsunami Information Center (under the auspices of the United Nations Educational, Scientific, and Cultural Organization-Intergovernmental Oceanographic Commission) and as chief scientist for various missions sponsored by the United Nations development program. When we spoke he told me that a search of his databases also failed to locate a tsunami occurring on August 29, 1916, in an area that would impact Santo Domingo. However, several hurricanes *had* occurred in the Caribbean on that date and on days immediately preceding. One went west to hit Yucatan; another curved northwest and hit Corpus Christi, Texas. A tropical depression arose on August 27 at latitude 46 degrees west, longitude 14 degrees north. It subsequently became a category one hurricane and, by August 29, a category two. At 12:00 GMT (4:00 P.M. Santo Domingo time) it was located at latitude 15.6 degrees north, longitude 67.6 degrees west. As it traveled west, the counterclockwise pattern of winds (as viewed from above) would send swells to the northwest, in the direction of Santo Domingo. Waves would build and move—the faster, long-period waves arriving first. Given the early storms, these waves could have encountered others and might have added to create the three large successive waves (a phenomenon now known as the "Three Sisters") that hit the *Memphis*. In summary, Pararas-Carayannis stated that he doubted that a tsunami caused the loss of the vessel.[17]

Regardless of the source of the wave, it is clear that several waves—one of which was at least 70 feet high—hit the *Memphis*, rolled her, lifted her up, dropped her down on the bottom of the bay, and surfed her to ruin on the beach about a half-mile distant. This is something to remember when selecting an anchoring spot.

4

Tempests and Storm-Tossed Seas

Mariners try to avoid storms and the rough seas that accompany them. Sometimes, however, that just isn't possible. One spring, shortly after I'd purchased *Dreams*, Nancy and I took our good friends Mark Reedy and Valerie Clarke to Catalina Island. They were both from the Midwest, caught in the throes of a particularly long and miserable winter, and looking forward to spending some time in sunny California. We sailed to the isthmus of Catalina Island, where we picked up a mooring and planned to spend several days hiking and exploring the island. Unfortunately, the weather took a sudden turn for the worse, and by Sunday morning, our planned day of departure, a small-craft advisory was issued, warning of winds of 30 knots and gusts of 35 or 40 knots. Since we all had travel plans for the coming week, we could not stay and sit out the storm. One option was to return to the mainland on the *Catalina Express*, a fast, twin-hull passenger ship destined for the Port of Los Angeles. But if I did this, it would mean leaving my new boat unmanned on a mooring in Isthmus Cove, Catalina, and there was no way I would do that.

Thus, we decided that Nancy and Valerie would return to the mainland on the *Catalina Express*, take a taxi from the Port of Los Angeles to

Marina Del Rey, which was my home port at that time, and meet us there. Mark bravely agreed to help me sail back to Marina Del Rey. *Dreams* is a sturdy boat, designed for blue water sailing, with a full keel, high shear on the bow, big drains in the cockpit—everything needed for heavy weather. I wasn't worried about the boat not being equal to the task. Mark and I got under way soon after that, one reef in the mainsail. Leaving the island, we saw that conditions really weren't so bad, so we put up the staysail along with the mainsail and put out a fishing line. All was well for the first hour or so. After that, once we were in the San Pedro Channel and away from the lee of Catalina, the seas and the wind started building. The fishing line came in first, then the staysail was furled. As we got closer to the mainland, the wind—originally out of the northwest—swung around to the north, and soon we had both wind and waves "on the nose" as the saying goes, meaning we were going directly into both wind and wave.

At this point, Mark was hunkered down in the cockpit, feet braced against the bulkhead, holding on to the cockpit rails. We both had on lifejackets and safety lines. We were also both wet and getting wetter. I had to impose on Mark to take the wheel so I could go forward and drop the mainsail—a task that he did not welcome, but he nonetheless acquitted himself well. For the next several hours we slogged our way north. After a while, I got the rhythm of the seas coming our way. I could maintain our northerly heading for six or seven waves and then we'd get a bigger one, and I would swing the boat to take it at a 30 degree angle to minimize the slamming. This worked most of the time, but not always; sometimes a second or third big wave followed the first. With the biggest waves—around 10 feet high—water would break over the bow pulpit 8 feet above the waterline and come back amidships before running off. Seawater would fly clear back over the dodger into the cockpit, drenching us as if a giant hand had thrown a trash can full of cold seawater in our faces.

For those who are not familiar with the Marina Del Rey harbor, the main channel runs northeast-southwest and has two entrances, one on the northwest side and one on the southeast side, the latter our direction of approach. There was a lot of shoaling at the south entrance at this time. In front of the harbor entrance there is a long breakwater running parallel to the coast. Boats enter between this breakwater and the jetties that form the edges of the main channel. As we approached the southern entrance, I could see that good-sized seas were running down the coast, into the north entrance and out the south entrance. I informed Mark that the trip was going to be a little longer—I didn't want to go into the south entrance with those seas running. I could imagine getting in that confined space and having the engine die or something else happen. So, we beat our way north past the breakwater and then came in the north entrance, bare poles, motor idling—at 7 knots, the fastest the boat had gone since I'd owned it!

Mark and I are still good friends, but if you ask him about sailing, he will roll his eyes and say, "Let me tell you about the time. . . ." As in many such incidents, the waves get bigger with each telling of the story. Afterwards, when we cleaned the boat up, it was amazing to see how much salt there was on it—everywhere, in every nook and crevice.

As storms go, this one was puny. Later we will hear from sailors with considerably greater experience, who have braved seas two to five times higher than these, and a few who have seen seas 10 times as high—100-foot extreme waves!

STORM ORIGINS

The storm we experienced was not severe and was typical of those occurring during the winter and early spring months in Southern California. A common cause of such storms in our locality is wind shifts that bring cold air from Canada or Alaska south into California, as opposed to the more typical track southeast into Montana, Wyoming, and Colorado.

The leading edge of a mass of cold air, moving fast at around 600 miles per day, is called a *cold front*. When cold dense air meets warmer air, it slides under and pushes the lighter, warm, moist air upwards. As the warm air rises and cools, its temperature drops, and moisture condenses, first forming clouds and then rain as the air saturates. *Warm fronts* are the leading edges of warm air masses that typically move more slowly than cold fronts (180 to 360 miles per day). In the United States, they might originate in the Gulf of Mexico, or in the Pacific or

Atlantic along the Tropic of Cancer.[1] Fronts arise in response to some disturbance or force aloft. The appearance of a front is usually accompanied by a drop in atmospheric pressure. The pressure drop can be gradual or rapid; if it is rapid and prolonged, the resulting storm will be severe.

Meanwhile, the entire system generally moves in an easterly direction, driven by the prevailing winds. Rain and strong winds can occur at the interface between the two fronts but also can occur elsewhere. For example, the most important weather with a traveling low-pressure system is the warm sector to the south and east of the low, where the rising air and typically strong south-southwest winds contain the most active showers and thunderstorms. Such storms are sometimes called *extratropical cyclones* since they have the characteristic rotating motion of a cyclone, but being outside tropical waters lack the driving force of the warm water "heat engine" that creates a true hurricane.[2]

AROUND THE WORLD ALONE

One of my heroes is Brad Van Liew. Brad is modest, unassuming, and one of the handful of people in the world who have sailed around the world alone in a small boat. He is probably the most talented sailor I know. Consequently, I knew I had to go to Charleston—his home base—and talk to him about storms and waves for this book. A word about Brad: He started sailing at age 5; beginning at age 12 he went to Newport, Rhode Island, during the summers to work on boats for his uncle. As a teenager, he gained sailing experience by working as a crew member during the Newport to Bermuda race, the Newport to Annapolis race, and other offshore regattas. He is also an experienced aircraft pilot, with multiengine, instrument, and instructor's ratings. For several years after he graduated from the University of Southern California, he operated an aircraft charter service.

Brad first considered taking part in the Around the World Alone Race in 1990, while still in college. He took leave from the university during his junior year and tried to raise the money to get a boat and enter the race, but he was unable to get the financial backing he needed. In 1996, 28 years old and now married to Meaghan, he decided to try again, this time for the 1998-1999 race. The race would depart from Charleston, South Carolina, and continue to Cape Town, South Africa; then from Cape Town to Auckland, New Zealand; from Auckland to Punta del Este, Uruguay; and from Punta del Este back to Charleston—a total distance of 25,400 nautical miles. Meaghan was an invaluable ally, helping raise the money needed for the race and also managing his support team. This time, with support from friends, borrowed money, and several corporate sponsors, he was successful.

You sail alone in this race. There is no one to stand watch at night so you can sleep. If something breaks, you are the one to fix it. Not only do you not have a crew, but since most of the race takes place in some of the world's most distant and inhospitable waters, you are unlikely to even see another vessel. Brad's boat, named for his principal sponsor, was *Balance Bar*, an open 50 class sailboat, 50 feet long on the deck, beam of 14.5 feet. This was his home at sea for nearly five months through some of the roughest oceans known to man. It is best to let him tell the story in his own words.

"I was approaching Cape Horn and was around 700 nautical miles west of the Cape, when I got word that a major depression was forming. I knew about a week beforehand that it was coming. There was a free fall of the barometer as the storm approached from the west—it dropped to around 920 millibars, as I recall. As the storm approached, winds were at 70 knots, and swells heaped up from the northwest with 20-foot-high waves. [Note: when speaking of wave height, Brad is using wave *amplitude*, not crest-to-trough. Crest-to-trough for a 20-foot-high wave would be 40 feet.] As the storm passed the boat, the wind became southwesterly, gusting to 100 knots, and the swell direction changed, with the swells now 30 feet and sometimes running into the swells coming from the northwest. The problem with multiple swells is that they can start colliding and at some point if you're in the wrong place, they'll grab you and spit you out the top."

At this point in our interview, in a courtyard near the swimming pool outside the hotel where I was staying in Charleston, South Carolina, Brad looked around and directed my attention to the hotel building behind us.

"When I was in a trough, the oncoming waves were as high as

that," he said, pointing to the building. I looked—the wall of the building loomed over us. It was five stories high.

"*That* high?" I asked.

"Yes—probably 30 feet or so—equivalent to 60 feet crest-to-trough. The top of my mast was 75 feet above the deck, and the waves were as high or higher.[3]

"Anyway, once the wind clocked around to where it was blowing to the southwest, it was blowing against those northwesterly swells, causing the faces to get steeper. At this point I was down below, braced at my navigation table in the center of the boat. We were running on autopilot, bare poles, and *Balance Bar* was surfing those big swells at 12 knots due east toward Cape Horn. Suddenly the boat heeled over 90-plus degrees so the mast was parallel to the water, and the boat slid sideways down the face of the wave. I happened to look at the global positioning satellite instrument at that moment and saw that my direction of travel was now south and *Balance Bar* was making 15 knots sliding down the swell on the boat's side.

"I said to myself, 'Come on baby, come on,' and moments later *Balance Bar* righted herself and the mad dash continued. Something was banging around on deck—the spinnaker pole had broken lose. At this point I decided I'd better go topside and see what things looked like. On deck, with foul weather gear, goggles, safety line, freezing water, snow, and hail—you couldn't see without goggles to protect your eyes, the wind was blowing so hard—it was an amazing sight. I fixed the spinnaker pole and then spent about an hour on deck watching—I was mesmerized by the seas—I'd never seen anything like them before.

"On most waves, the boat would ride up and the swell would pass under the boat and break later. Every now and then a larger one would come, maybe 40 to 50 feet, with the classic 'graybeard appearance.' The top 3 to 6 feet would be foamy streaming grayish water, churned and blown off by the wind. Below this was the dangerous stuff, a wall of water that could break the boat or crush you. As I watched, I could see the surface layer of this wall of water sort of break loose and slide down the face with a glistening, white rippling effect, resembling water sliding down a water slide in an amusement park ride. Finally one of these waves broke on the boat and we began the down-wave slide again, me hanging on for my life as tons of water poured over me and over the boat—holding my breath as long as I could until finally the water ran off and the boat righted. At that point I decided I'd better get back down below.

"I got into my 'coffin bunk,' so-called because it was made so I could squeeze into it and not get tossed around when the boat's movements got pretty wild. Things were fine for a while and then all of a sudden it got dark and very quiet, and I found myself on the overhead—the roof of the boat. There was a plastic bubble above my navigation station where I could look out and see how we were doing. Looking into this, all I could see was very deep blue water. I realized that the boat had broached; I was upside down in the Southern Ocean. It was an eerie feeling. At one moment there was the howling of the wind, the rattling and banging sounds of water hitting the deck, wind whistling through the rig—then suddenly, silence. Being upside down made for quiet running. The light was surreal—no more white light from the sky, but diffuse blue light—daylight being filtered through the ocean, then coming in the windows that were underwater.

"My first thought was, 'Well, this is it. I'm going to die.' Then I got angry—angry at the sea, angry at myself for putting myself in the position, mad about a lot of stuff. About then another wave hit the keel, gave *Balance Bar* a nudge, and we rolled around and came upright. After a few moments to calm and steady myself, I went topside to survey the damage. Amazingly, the rig and mainsail were still there—no major losses. There were some minor things, but I could fix them. Later I found out that the mast stays and standing rigging were badly stretched and I no longer had a nice tight rig.

"I had four or five more knockdowns, but didn't roll again, and managed to clear Cape Horn and get into Punta del Este, a scheduled four-week stopover, where Meaghan and my support team were waiting for me. We had some discussions about whether it was worth it— whether I should continue or not. Meaghan and I had previously agreed that if either of us decided it was time to quit, I would quit. But we decided that, having come this far, I would continue for the last leg back to Charleston. Besides, wasn't the worst of it behind me? Hadn't I made it around the Cape?

"I should have replaced the rig, but without money the best I could do was tighten everything up—retune it—and hope for the best. With this I left Punta del Este for the last leg home. Leaving the Rio Plata, near the Uruguay-Brazil border, I stayed inshore. There is a continental shelf there. This was a mistake—I should have immediately headed for deeper water. As it was, I got into some really rough water—waves that were only 6 to 8 feet high, but they had short wavelengths and really tossed the boat around, raising it and then dropping it with a crash. The strain was too much for my beat-up rig and the mast broke in three places. At that point I was really demoralized and ready to call it quits, but when I called Meaghan on my satellite phone she and the crew urged me not to quit.

"I jury-rigged my spinnaker pole as a mast and got up two sails—my storm jib and another—and limped back into Punta del Este. It took me two days to get back and then seven days for repairs. Thanks to Meaghan's efforts and the help of many other people, *Balance Bar* got fitted with a new mast and I restarted the race, 1,000 nautical miles behind the other competitors. With good luck and hard sailing, I managed to catch up with the fleet and finished third, only a couple of weeks behind the leaders.

"So, here's a lesson for your book; waves don't have to be big to be mean."

There is a sequel to this story. Van Liew competed again, in the 2002-2003 race, this time in a boat called *Freedom America*. He not only won the race in his class but completed an unprecedented sweep of all five legs. I asked Brad how this race compared to his first attempt.

"On the second race, the southern oceans were not the toughest part of the race," he said. "I went from one low to the next. The lows march from west to east across the Southern Ocean, moving at about 30 knots. I tried to keep on the north side of the low, where I consistently had 15- to 20-foot waves and 25- to 40-knot winds. These waves were 200 to 300 yards apart (about 15 to 20 boat lengths), so I'd ride up on one and then I could see the last one out there ahead of me. I like to stay on a beam reach to a broad reach. I found that I could slalom through those swells. I was able to maintain a fast speed—averaged around 9.5 knots over the entire 7,800 miles sailed between New

Zealand and Salvador Brazil. When the wind gets above 40 knots, the waves start to heap up and you can't go as fast.

"I found the Indian Ocean to be the most dangerous. Leaving the Cape of Good Hope, you encounter the Agulhas Current. In a previous race, two boats rolled in the area. One—I think it was a modified Swan 44—pitchpoled two times. You want to get past the Cape and into deeper water. However, heading east towards New Zealand, you first pass the Crozet Islands and then the Kerguelen Islands. These lie in shallow waters and you can have some horrendous seas building there."

After 148 days at sea alone, Brad triumphantly sailed into Newport, Rhode Island, and into sailing history. He had arrived a full three weeks ahead of his nearest competitor.

AN AMAZING RESCUE

On the west coast of Alaska, south of Valdez, destination of oil tankers, is a mountain called Mount Fairweather. Not far from it is a beautiful, secluded bay called Lituya; more about that later. This is the area swept by the Alaska Current as it curls northwest from Vancouver and then streams west along the Aleutian Island chain. It is an area of unsurpassed fishing grounds, prolific in its production of salmon and other species. It is also unsurpassed in the spawning of storms characterized by high winds and giant waves. It became the site of an incredible rescue in horrendous seas.

Fishing is a dangerous occupation—certainly in these waters, far more dangerous than coal mining or almost any other hazardous employment. At the same time, the area between Sitka and Anchorage is one of spectacular natural beauty—sharply rising mountain ranges, crystalline blue water beneath the edges of glaciers, bald eagles perched in treetops overlooking isolated coves. During the summer months, you would be hard-pressed to find any area of the world more beautiful, and during the winter months, any place with deadlier weather. Dense, cold air created near the tops of the high mountains along the coast settles down long valleys toward the sea, creating sudden, unpredictable, tumultuous winds—williwaws—that blast out into the ocean. Storms blow in from the Bering Sea or arise in the Gulf of Alaska, creating a nightmare for boats large and small. Storm winds frequently reach Beaufort Force 12 at 80 to 110 knots and, blowing over a long fetch of up to 430 nautical miles, can produce 80- to 100-foot-high waves.

Salmon trollers and others who make their livelihood from the sea depend on the United States Coast Guard as their ultimate lifeline in times of disaster. For this area, the Coast Guard operates from a base in Sitka and has a main support center in Kodiak. Here, one of the most amazing air-sea rescues of all times took place in January 1998. Late in the season, the fishing vessel *La Conte* went out to Fairweather Ground. In this location, about 60 nautical miles from Cape Fairweather, the bottom rises up to within 13 fathoms of the surface and there is excellent fishing for red snapper and other bottom fish, although January is a dicey time to venture out to Fairweather Ground.

The *La Conte*, with a crew of five, got caught in a horrific storm and started sinking. Spike Walker, a writer, and a man who has fished in these same waters, interviewed the surviving crew members and the Coast Guard personnel who rescued them, and wrote a riveting account of what happened to them in his book *Coming Back Alive*.[4] Just before *La Conte* was finally pushed under by a huge wave, the crew was able to get into their insulated survival suits and grab the emergency position indication radio beacon (EPIRB). Tying themselves together with a length of line, they jumped into the frigid water. Their only thread of hope for survival was the EPIRB. Fortunately, they had the newer type that broadcasts on 406 megahertz.

Dreams is equipped with a similar instrument, but other than periodically testing it to make sure it is functional and the batteries are still good (they are supposed to have a five-year life), I have never had to use it. It will activate automatically if placed in water or can be turned on by a manual switch. Each EPIRB has a unique identification number that is registered with the National Oceanic and Atmospheric Administration. This provides information about the vessel: size, color, radios on board, and other information essential to a rescue operation. Once the EPIRB is activated, it sends a 406-megahertz signal to one of several overhead satellites (the COSPAS-SARSAT network). The signal enables the satellite to determine the latitude and longitude of the

EPIRB and to alert the nearest Rescue Coordination Center. The EPIRB also transmits a homing signal on 121.5 megahertz so that as a rescue vessel or aircraft gets close, it can home in on the exact location. A high-intensity xenon strobe light flashes to make visual identification possible at night. The battery is designed to last 48 hours.

Within minutes, notification that an EPIRB had been activated at Fairweather Ground reached the U.S. Coast Guard base at Sitka. A Coast Guard helicopter was dispatched to the area, which was about 110 nautical miles northwest of Sitka. Miraculously, the helicopter found the tiny dot of flashing light in mountainous seas—and the four men tied together. (The fifth man had been pulled out of the rope by a large wave and disappeared from sight, presumed drowned.) The pilot tried to maneuver the helicopter into position above the men, fighting 70-knot headwinds with gusts that were even higher. Time and time again as he descended to 100 feet above the sea to deploy a lifting basket, he had to pull up when a huge wave threatened to knock the helicopter out of the sky. When he pulled up, the fierce winds blew the helicopter backwards, sometimes as much as one-half mile, and he had to fight his way back and relocate the crew. Once, flying at an altitude of 100 feet, as the pilot concentrated on maneuvering the helicopter, a crewman suddenly screamed, "Up, up." As he looked out the door in preparation for lowering the lifting basket, he looked *up* and saw a rogue wave higher than the helicopter about to break on them. With engines straining at maximum power and skillful flying, they escaped by the narrowest of margins. Back and forth the battle with wind and waves went, until the helicopter reached the point of no return on fuel and had to return to Sitka. Below, the men in the sea struggled to keep their hopes up, to keep together, and to fight their way back to the surface after each 70-foot-high wave buried them under tons of water.

A second helicopter was dispatched from Sitka, but the results were the same. Despite numerous heroic efforts, it could not maneuver the rescue basket close enough to the men in the water to retrieve them. A third helicopter was dispatched. By now, the men had been in the frigid waters for hours, constantly battered by waves 70 to 90 feet high, and were slowly losing what little strength they possessed to assist in their own rescue. Finally, after nearly seven hours in the water, the third helicopter managed to place the rescue basket about 30 feet from the stranded fishermen. One man cut himself loose, grabbed a second man, and swam to the basket. He managed to get in, but the second man hung half in, half out, and as the helicopter crew frantically tried to raise the basket, it was slammed several times by giant waves. Finally up, the crew succeeded in getting one man into the helicopter, but at the last minute the second man lost his grip and fell to his death. Again the basket went down, again two crewmen got in, but at that moment a huge wave crashed on the basket and knocked the second man out. Now the third helicopter was running low on fuel, but the crew elected to try a couple of times more and miraculously got the basket close enough to the third survivor that he half swam, half floated into it and was saved. Later, the fourth crewman was recovered by yet another Coast Guard helicopter, this time sending a diver into the waters to recover the body.

The helicopter crew members who rescued the *La Conte* survivors were awarded the Distinguished Flying Cross for their efforts. Sometime, long ago, in the midst of launching a small rescue boat in heavy seas at Cape Hatteras, a Coast Guard sailor looked at the conditions and lamented, "We may make it out, but we'll never make it back," or words to that effect. The man in charge replied, "The regulations say we've got to go out, but they don't anything about having to come back." The phrase "You have to go out, but you don't have to come back" has become a watchword for U.S. Coast Guard response to marine emergencies.[5] Anyone who spends time on the water has only the greatest respect for those who risk their lives daily so that others might live.

SQUALLS

A *squall line* sometimes can be observed in advance of an approaching front, but more often than not, there is no front nearby. A squall line over the ocean at higher latitudes may be due to the outflow of cold air pushing ahead and causing a boundary. It can range from 20 to 1,000 miles in length. A squall is created when there is a flow of moist air in front of the advancing cold air. The denser cold air causes the warm moist air to rise. As the system approaches, sometimes moving swiftly, there will be a sudden onset of higher winds and precipitation. As it comes nearer and passes overhead, you can see a dark black cumulus cloud mass. Wind and rain are intense but usually brief. Squall lines can be accompanied by thunderstorms.

Squalls have been life savers to persons stranded at sea without water. My favorite story of survival at sea concerns three young navy fliers who had to ditch their carrier-based scout bomber in the South Pacific in the early days of the Second World War. Their story impressed me more than most of the other great survival epics (Captains Bligh and Shackleton, for example) because in the other cases they were experienced mariners, with seaworthy boats, food, water, and navigation instruments, while the navy crew had no experience, no equipment to speak of, and only their sheer determination to survive.

Barely in their twenties, two were from the midwestern United States and had no special knowledge of oceans other than the introduction they got in basic training. The pilot, however, lived near San Diego and had grown up near the sea. The three of them were thrown together as a crew for this particular flight; they did not even know each other well. Of the three, only Harold Dixon, the pilot, knew how to swim. Yet with ingenuity, great courage, and resolution, Dixon guided their small 8-foot-long rubber raft for more than 900 nautical miles and 34 days to a successful landing on a small inhabited South Pacific island, where they were eventually rescued. To accomplish this, they had a 0.45-caliber automatic pistol and three clips of ammunition, a pair of pliers, a pocketknife, a whistle, a small mirror, a fabric bailing bucket, a manila lifeline, two life jackets, a pencil, and the clothes they were wearing. Dixon used the pencil and one life jacket to construct a chart of the area from memory and, from the last known position of the plane before it crashed, determined a course to the nearest island. He improvised a sea anchor with the line and the other life jacket, improvised paddles from a pair of shoes, kept a course by dead reckoning, and, without oars or a sail, "sailed" the life raft to a safe landing. By the time they landed, they had lost all of their equipment in rough seas during the numerous times the raft swamped and overturned, as well as all of their clothes. The only thing that remained was the whistle.

They survived by catching rainwater from squalls, spearing fish with the pocketknife (including one good-sized shark), and catching or shooting birds (before the pistol became too rusted to use). A few floating coconuts made up the rest of their food supply. When a squall approached, they removed their clothing and used it to soak up water, which they first drank, storing any extra in the bailing bucket.

Once ashore, friendly natives fed them and restored them to health. For the first week on land, they were unable to walk. Shortly after landing on the island, the barometer started falling and winds built to gale force, and then became a full-blown hurricane that lasted three days. Great damage was done to the island: About one-third of the trees were blown down, and taro and other crops on low-lying sections of land were wiped out as great waves washed inland and over parts of the island. No one wants to imagine what would have happened had they still been in their tiny raft when the storm struck.

Dixon subsequently received the Navy Cross for "extraordinary heroism, exceptional determination, resourcefulness, skilled seamanship, excellent judgment, and highest quality of leadership."[6]

NORTH ATLANTIC STORMS AND WAVES

The North Atlantic Ocean has historically been noted for horrible winter storms. Even the biggest ships in the U.S. Navy are not immune to the effects of large waves. One of the first supercarriers, the USS *Forrestal* (CVA-59), 1,076 feet long, saw its share of rough weather during 21 deployments between 1954 and 1993. My good friend Ray Holdsworth was a young lieutenant (junior grade) on the *Forrestal* from 1965 to 1967.

Ray described to me the winter of 1965-1966, when the carrier returned from a Mediterranean cruise to Norfolk, Virginia. The *Forrestal* passed Gibraltar on its way into the North Atlantic. A few hundred miles south of Newfoundland, a large storm arrived, bearing down on the carrier from the northeast. The storm lasted for one and one-half days and left in its wake enormous swells—too much for even the mighty carrier to take head-on. The *Forrestal* altered its heading to take the swells at 15 degrees and slowed to 8 knots. Even then, green water flowed over the bow and onto the flight deck 60 feet above the waterline.

Normally, rough weather had little impact on the carrier. It was the destroyers and other escort vessels that suffered. During really rough seas, serving meals was impossible on the smaller vessels and the crew made do with snacks or whatever they could grab as their vessels pitched and rolled. The carrier would alter course at the dinner hour so the escorts could prepare and serve an evening meal. This was known as a "dinner course."

In June 1967, the *Forrestal* was deployed to Vietnam. Shortly before 11:00 A.M. on July 29, 1967, as *Forrestal* was preparing to launch planes in the Gulf of Tonkin, a fire broke out on the flight deck. Planes blazed out of control, and bombs and ordnance exploded. Ray was climbing a ladder leading to the bridge at the time; the blast literally knocked him off the ladder. When the fires were finally put out, more than 130 crewmen were dead and hundreds injured in the worst naval disaster since World War II.

After making emergency repairs, the *Forrestal* limped into the big naval base at Subic Bay, Philippines, for additional tests and repairs. From Subic Bay she steamed west through the Sunda Strait between Java and Sumatra. Crossing the Indian Ocean, her route took her past Madagascar, where she rendezvoused with a British tanker, and then around the Cape of Good Hope and back to Norfolk and dry dock at the Portsmouth naval shipyard for repairs.

The *Forrestal* traveled alone—the other ships could not be spared and remained behind in Vietnamese waters. All her aircraft were removed (those that had not been destroyed), since flight deck damage precluded launching and recovering aircraft. So, unarmed, unable to defend herself, unescorted, with major holes in the flight deck (some 20 feet in diameter), the *Forrestal* passed through the Agulhas Current, rounded the Cape, and entered the South Atlantic. Fortunately the Agulhas Current remained calm, the weather was benign, and no more problems were encountered. Ray remembers it as a long trip of 34 days; the maximum speed the *Forrestal* was capable of at that time was 12 knots, and the loss of so many crewmates had saddened the entire crew.

If any vessel is impervious to heavy weather, you'd think it would be a nuclear-powered aircraft carrier. I asked Rear Admiral Bill Cross, a U.S. Naval Academy graduate who commanded the nuclear aircraft carrier *Dwight D. Eisenhower* from 1990 to 1993 about carriers and heavy weather. The *Eisenhower* is 1,092 feet long and carries a wing of 70 aircraft. The flight deck is composed of steel plates 1.5 inches thick. The crew totals approximately 5,800 persons, of whom 3,000 operate the carrier and 2,800 operate and maintain aircraft.

In the summer of 1991, the *Eisenhower* was back in the Persian Gulf following the first Gulf War. One of Admiral Cross's vivid memories of that time was the Kuwait oil well fires. With more than 500 oil wells burning, smoke rose high in the sky, where the prevailing winds spread it out in a huge black layer. At night the flames from all of the burning wells lit up the night sky, turning the underside of the immense black cloud orange and red, visible even to the carrier far out in the Gulf, and making the demanding task of flying from a carrier at night even more difficult.

The *Eisenhower* next was deployed to the North Atlantic to take part in a North Atlantic Treaty Organization exercise in March 1992. The exercises were planned to take place off of Norway. Word came of a large storm coming up through the North Atlantic. As the storm approached, the air pressure dropped to 920 millibars and winds were 70 knots. All night, as the carrier raced north at 34 knots, the seas on the port quarter imparted a rolling motion to the carrier. Waves were typically 40 feet, although some were 60-plus feet because they broke over the bow of the carrier and washed over the flight deck. The carrier was rolling at about 6 to 8 degrees. This was hazardous because there were approximately 50 aircraft chained down on the deck. The cyclic stress of the rolling motion tends to loosen the chains: too loose, and a $40 million aircraft would roll over the side. Crewmen had to venture out onto wet, icy decks, in the face of intense winds, to check and periodically tighten the chains. Whaleboats—secured on the side of the vessel, high up in sheltered mounts—were stripped away and lost by the force of the storm. Further inspection of the vessel showed that the hurricane bow (steel plates at the bow of the vessel below the flight deck) had been bent inward by the force of the waves. Temporary repairs were made by welding steel I-beams to the deck inside to brace and reinforce this area of the ship. The storm finally abated as the *Eisenhower* reached the coast of Norway.

THUNDERSTORMS, LIGHTNING, WATER SPOUTS

Weather conditions are key to understanding how extreme waves on the surface of the ocean form and propagate. The next several pages summarize some of the principal types of weather likely to be encountered at sea.

Thunderstorms occur when there are strong upward currents of warm air with considerable water vapor. The air rises to an altitude of a few miles to as much as 12 miles, forming the distinctive cumulonimbus cloud rising to an "anvil" top. The anvil appearance is caused by the high-velocity winds in the upper atmosphere blowing the tops away from the updraft. The rising air reaches an altitude at which the temperature is well below freezing, causing small particles of ice to form. In the center of this turbulent region of swirling air and abrupt temperature, electrons are stripped from some atoms and are swept to other parts of the cloud, leaving some portions negatively charged and other parts positively charged. When the potential difference has built to a high enough value to break down the resistance of the air, a lightning flash occurs. Thunder is the sound of lightning; the rapid passage through the air of the lightning flash's huge electrical current heats air to a high temperature and creates a shock wave we know as thunder.

Waterspouts are caused by a strong updraft over a large body of water. They are more likely to occur in tropical latitudes where the water temperature is at least 27 degrees Celsius (80 degrees Fahrenheit) than in higher latitudes. Lower atmospheric pressure within the center of the rotating column of air sucks water from the surface of the sea and also condenses water vapor in the swirling air. A waterspout is similar to a tornado but occurs over water and can be as high as 1,500 feet and last as long as half an hour. Waterspouts occur occasionally in the Pacific near Southern California as well. One or two per year are observed here, usually close to land. While they can be damaging to small craft, it is usually possible to avoid them. However, in December 1969, a waterspout struck the Huntington Beach, California, pier, destroying the head of the pier, injuring 17 people, and killing three.[7]

EXTRATROPICAL CYCLONES

Large storms also arise in the middle and high latitudes where the ocean water is cooler than 25 degrees Celsius (77 degrees Fahrenheit). Warm air flows mainly horizontally toward the low, and there is a lesser but important component that is rising near and within the low. Cold air also flows mainly horizontally toward the low and tends to sink along the way. The occluded front, when there is one, is the boundary between the patterns. At the ocean surface, there are clear boundaries between the air masses along the cold, warm, and occluded fronts, while aloft there are not many sharp differences. Such storms are caused by cold polar air. The denser cold air overtakes the warm front of the depression, creating an occluded front where the warm air is forced upwards by colder air (occluded means "blocked"). Sometimes these occlusions fade away, but at other times they grow into huge rotating storm systems, as great as 3,000 miles in diameter. Extratropical storms differ from hurricanes; there is no warm, clear eye. Also—and most important—is that they have a large area with fairly similar, widespread, strong winds and waves. The storm moves in an easterly direction, traveling at a little more than 90 degrees to the occluded front, which is the boundary between the warm sector and the cold region. An abrupt wind shift often marks the passage of the occlusion, resulting in steep, confused seas.

The barometric pressure drops associated with extratropical storms are not as great as those of hurricanes, and the maximum winds are often less powerful than those of major hurricanes. The great danger from extratropical storms arises from the fact that they can extend over a much broader area and can last for several days. With longer duration and greater fetch, they can produce very large waves and dangerous seas. The resulting swell can travel extraordinary distances, as noted in Chapter 5. A rapid fall of the barometer is a sure sign of an intense storm; however a steady barometer is not a guarantee of good weather. Many a storm has seemingly "come out of nowhere" with no barometric warning.

TROPICAL CYCLONES, HURRICANES, AND TYPHOONS

A more serious storm is a *tropical cyclone*. These are storms with winds that rotate around a low-pressure area as described above but require warm water (26 degrees Celsius, 79 degrees Fahrenheit) as one of the driving forces. In their most violent form—when their winds exceed 64 knots—they are called *hurricanes* in the Atlantic and East Pacific and *typhoons* in the Northwest Pacific. These names have an interesting etymology. Hurricane is derived from Hurican, the Carib god of evil, and was probably conveyed to Columbus or one of the later Spanish navigators who followed him. Hurican may himself be patterned after the Mayan god Hurakan, who was believed to have taken part in the creation and whose breath created storms and floods. Typhoon probably came from the Chinese *tai fung*, meaning big wind, or possibly from the Greek *tuphōn*, after the Greek god Typhon, a mythological monster with many heads believed to be the source of whirlwinds and hurricanes.

However, to be classed as a hurricane, a tropical cyclone must also have an eye, an eyewall, and outer feeder bands that spiral into the center. In the higher latitudes, these features are mostly absent. In a North Atlantic or North Pacific cyclone, there may be a clear center, but it is diffuse and broad, not tight like a tropical hurricane. The difference is usually obvious from satellite and radar imagery.

Tropical cyclones always begin as *tropical depressions*, a name given to storms in which the winds are flowing in a circle around the low-pressure zone but have speeds less than 33 knots. The name derives from the characteristic low barometric pressure at the center of the storm, which is "depressed" compared to the general area around it on the weather map. Sometimes a tropical depression remains just that; it moves into an area where conditions above it and around it cause it to weaken.

Alternatively, a tropical depression can gather energy and grow. Meteorologists have identified some of the conditions necessary for a tropical depression to become a storm. The first requirement is warm water—at least 26 degrees Celsius (79 degrees Fahrenheit)—covering several hundred square miles of ocean or more. This pool of warm water needs to be at least 200 feet deep, otherwise the turbulence from an expanding storm could stir the water enough to bring up colder water and shut down the process. When conditions are right, warm, moist air flows into the tropical storm near the surface, following a spiral path inward. It speeds up as it approaches the center. As it rises, the air cools, some of the water vapor condenses into small drops of water or ice crystals, forming clouds and rain and *releasing heat.* This is a critical feature of a tropical storm, the reason it is sometimes called a heat engine. At the upper part of the storm (say, above 20,000 feet) the air spirals outward. The rising warm air creates a low-pressure area that causes still more air to flow inward. As the process expands, the wind near the sea's surface strips water droplets and spray, increasing the humidity in the column of rising warm air and adding more heat energy to the process. Finally, if the winds in the vicinity of the growing storm are light or maintain the same direction and speed to an elevation of around 40,000 feet, the nascent storm will continue to grow, its winds accelerating until it becomes a tropical storm with wind speeds in the range of 34 to 63 knots. When the sustained winds reach or exceed 64 knots, the storm is called a hurricane or a typhoon. At this point it is moving in a generally westerly direction at 5 to 15 knots. It is characterized by a central region called an eye that is 6 to 38 miles in diameter and is surrounded by an eyewall; within the center there is little wind or rain. The air in the eye is warmer than the surroundings. In comparison to squalls and thunderstorms, tropical cyclones cover a wide area, have sustained movement along westerly-northwesterly tracks (usually), and can last for several days to a week or more.[8]

The passage of a hurricane has an interesting effect on the sea. After it passes, the surface seawater temperature drops by as much as 1 degree Celsius close to the center of the storm. In the first few days following a hurricane, the sea temperature along the hurricane's track may be as much as 3 to 5 degrees Celsius (6 to 9 degrees Fahrenheit) cooler.

Tropical storms can develop rapidly and move quickly, one of the reasons they were so dangerous to shipping from the time of Columbus up until the early twentieth century. Prior to the advent of satellites, radar, single sideband radio, and sophisticated weather forecasts, the mariner had to rely on his observations of the sea and sky. Gathering clouds, a change in swell direction, a falling barometer—these were potential warning signs. Knowing that a storm was coming was no guarantee of safety, because ship captains never knew for sure what direction the storm would take. Their ships were almost always too slow to outrun a storm, so unless there was a safe port nearby, they had no choice but to secure their vessels as best they could, shorten sail, and ride it out. Even today, with the latest weather and storm forecasting computer models, there is no guarantee that a given storm will behave as predicted.

Tropical storms originate in specific areas of the oceans where sea and wind conditions are suitable. They also generally occur at specific times of the year. For example, in the North Atlantic and the Caribbean—from Venezuela on the south to the East Coast of the United States and Central America on the west, Newfoundland on the north, and eastward to Africa—they occur from June 1 to November 30 and peak from August to October. The specific times for other major ocean basins are available in the literature.

The frequency of storms varies with location and from year to year. The western North Pacific is the most prolific source of tropical cyclones, averaging about 25 per year, approximately 18 of which become typhoons. Long-term records show that the frequency of hurricanes increases in some years and decreases in others, generally on a scale of decades. For example, 1991 to 1994 were "quiet years" in the North Atlantic and Caribbean, while post-1995 has seen an increase in activity. In Australia a downward trend has been observed, while in the Northwest Pacific the trend has been up. Likewise, the trend has been upward in the Northeast Pacific but downward in the North Indian Ocean. The reason for these changes is not understood, but is thought to be the result of shifts in ocean water temperature, the wind patterns in the upper atmosphere, and, in the case of the North Atlantic, the number and type of storms coming off Africa.

The paragraphs above should be considered with a caveat: Major storms can occur during any month in the hurricane season and sometimes (rarely) in other months. They also can sometimes originate, or travel, outside their usual boundaries. However, no hurricanes arise within 5 degrees north or south of the equator and a storm within 10 degrees is rare.[9]

The area of Southern California (latitude 33 degrees north) in which I live lies north of the main hurricane grounds of the Northeastern Pacific. Two popular cruising destinations from Southern California are the Hawaiian Islands or Baja California and the Sea of Cortez, Mexico. These trips need to be planned with hurricane avoidance in mind; for this reason, the annual Newport to Ensenada race takes place in April; the TransPacific Race, Los Angeles to Honolulu, in July. On the East Coast, the Charleston to Bermuda Race takes place in May.

Normally, hurricanes that are spawned in Mexican waters track out into the Pacific toward Hawaii and then dissipate—but not always. They can curve back and cross over Mexico or Southern California, eventually blowing themselves out over the California and Arizona deserts.

In the year 2000, to celebrate the new millennium, I planned to sail to Guadalupe Island (Mexico), a barren, windswept island 150 nautical miles west of the Baja California coast and 300 nautical miles south of Newport Harbor, California. I scheduled the trip near the end of October, hopefully past the storm season. The weather had not been great, but it finally cleared, and with two crew members—Russ Spencer and Erik Oistad—I got under way on a Friday night at 2130 hours (9:30 P.M.). It was a clear, moonless night, and it was a wonderful feeling to finally see the coast slip away behind us after days of preparation. *Dreams* sailed all night, finally reaching San Clemente Island at 8:00 A.M. We spent the next two days diving, relaxing, playing poker, and doing last-minute boat checks. Before jumping off for Guadalupe (a straight run to the south of several days and nights, depending on the winds), we made a final check on the weather, using National Weather Service weather faxes obtained over the single sideband radio. (See example in Chapter 3.)

Alas, the weather reports were not good.

The outlook had changed suddenly as we relaxed at anchor in Pyramid Cove. Cold weather and rain headed our way from Alaska, and there was a small-craft advisory of winds at 25 to 30 knots and seas of 10 to 18 feet. In the other direction, a tropical depression was coming up from Mexico; the seas around Guadalupe Island were forecast at 29 feet.

We decided to backtrack to Catalina Island and wait out the storms in Catalina Harbor, a run of about 40 nautical miles north, in the direction of the coming storm. When we left San Clemente Island at 9:00 A.M. it was a bright sunny morning; by noon the sun was gone and it was cloudy and overcast. By 2:00 P.M. we were in foul weather gear, it was pouring rain, and visibility was about 1 mile. By the time we reached the harbor at 5:00 P.M., rain was falling in sheets and visibility was down to a few hundred meters. We had to wait from Sunday to Tuesday for the weather to clear. Guadalupe was postponed until the next year, when we finally had perfect weather and a delightful trip to that island and three other island groups south of the border.

Normally, Southern California is spared the impact of tropical cyclones, but not always. I've sailed in or out of the Newport Harbor entrance at least 100 times, in the day and at night, in good weather and poor. There is a comforting feeling when returning at night from several weeks at sea or even from a weekend fishing trip, when the red and green lights marking the end of the two stone jetties that form the harbor entrance finally appear as faint dots on a dark horizon. At this point you know that the dock, home, and a bed that does not rock are not far away.

Not so in September 1939: It had been an unusually hot week—so hot that on Wednesday, September 17, schools were closed. By the weekend, people sought the beaches to get relief from the heat, and sailors took to their boats. Sunday, September 21, was cloudy and overcast. Shortly after noon the weather changed dramatically. In around 20 minutes, winds increased to 50 knots and soon were gusting to 65 knots. In the local area, such storms are often referred to by their Spanish name, *chubasco*—literally, storm, squall. When the storm hit, boats out sailing for the day struggled to get back into the shelter of the harbor. Huge waves slammed into the breakwaters at the entrance to the harbor, made even higher by collision with the outgoing tide. The storm impacted the coast as far north as Los Angeles Harbor, where towering waves lifted one sailboat over the breakwater and deposited it inside the harbor, and others knocked out the windows on the first and second floors of the San Pedro lighthouse.

Amazingly, the storm was documented by a bold photographer, who stood on the jetty shown in Figure 17 (Chapter 8) and filmed it with a movie camera. The film was placed in storage and forgotten for 50 years, but it was recovered in 1992 and incorporated into a documentary by the Newport Harbor Nautical Museum, where on a given day it is possible to watch dramatic jerky, grainy images as tragedy unfolds.

Huge waves can be seen rolling into the harbor. Their height is unknown, but because sailboats and fishing boats that appear on the crest of one wave totally drop from sight into the trough of the succeeding wave, they must have been in the range of at least 15 to 30 feet high. One sailboat races by, driven at hull speed like a giant surfboard. Next a 34-foot-long cabin cruiser appears, hesitates on the crest of one wave, gets hit by another wave, and capsizes with nine people on board. Eight are rescued—by a pair of local youths who ride out on surfboards! The ninth person, a woman, drowns.

The captain of another sailboat, returning from Catalina, wisely decides that returning to the harbor is not a good idea and elects to run before the storm. His boat is blown all the way to the Channel Islands, a distance of nearly 100 nautical miles, and returns several days later after the storm has died down. A power boat, the *Paragon*, 140 feet long, is returning from Catalina Island with 24 passengers and a professional crew at 8:00 P.M. It is now dark and the storm is at its peak. The captain refused to take the boat into the harbor, considering it too dangerous to do so. The owner then took over the wheel and tried to run the gauntlet of waves; the vessel was hurled sideways, hitting the jetty, which punched a hole in the stern quarter. Fortunately, the *Paragon* slid free of the breakwater and before she sank the owner managed to run her aground on a sandy beach at the end of the channel. Crew and passengers all survived; the vessel was later salvaged and repaired.[10]

In the aftermath of the storm it was learned that 45 persons lost their lives—in addition to those who were passengers in boats—23 in the Newport area. This was sufficient for the storm to merit a listing in the "Addendum" to the table of the "Thirty Deadliest Mainland United States Tropical Storms" that the National Hurricane Center maintains on its web site, which is reproduced later in this chapter as Table 3.

THE SAFFIR-SIMPSON HURRICANE SCALE

The strength of a hurricane is rated 1 to 5 on the Saffir-Simpson Hurricane Scale based on the hurricane's intensity at the time of reporting. The scale is used to give an estimate of the potential property damage and flooding expected along the coast from a hurricane landfall. Wind speed is the determining factor in the scale, as storm surge values are highly dependent on the slope of the continental shelf in the landfall region.

For example, a Category One hurricane has winds at 64 to 82 knots (74 to 95 miles per hour), a storm surge generally 4 to 5 feet above normal, and a central pressure less than 980 millibars. No real damage to building structures is expected in a Category One hurricane; damage is primarily to unanchored mobile homes, shrubbery, and trees, with some damage to poorly constructed signs and some coastal road flooding and minor pier damage. Hurricanes Danny of 1997 and Gaston of 2004 were Category One hurricanes at peak intensity.

At the other end of the Saffir-Simpson Hurricane Scale, a Category Five hurricane has winds greater than 135 knots (155 miles per hour), a storm surge generally greater than 18 feet above normal, and a central pressure less than 920 millibars. Complete roof failure on many residences and industrial buildings can be expected; some complete building failures with small utility buildings blown over or away; all shrubs, trees, and signs blown down; complete destruction of mobile homes; and severe and extensive window and door damage. Low-lying escape routes are cut by rising water three to five hours before arrival of the center of the hurricane. A Category Five causes major damage to lower floors of all structures located less than 15 feet above sea level and within 1,500 feet of the shoreline, and massive evacuation of residential areas on low ground within 5 to 10 miles of the shoreline may be required. Hurricane Mitch of 1998 was a Category Five hurricane at peak intensity over the western Caribbean. Hurricane Andrew of 1992 was a Category Five hurricane at peak intensity and was one of the strongest tropical cyclones ever to hit Florida.

HURRICANE-GENERATED WINDS AND WAVES

In the open ocean, the size of waves generated by a hurricane also depends on the force of the wind, the time that the wind has been blowing, and the condition of the sea (calm or stormy) prior to the arrival of the hurricane. However, hurricane-generated waves are different in one respect from ordinary wind waves—namely, the force producing the waves is moving in the same direction as the waves. For this reason the strength of hurricane-generated waves, especially in the long-period components, increases rapidly. Wind-generated waves do not always propagate in the direction of the wind.[11]

Waves produced by a hurricane depend on several factors, including the sustained wind speed, the pressure in the eye, the forward velocity of the center of the hurricane, and the radius of maximum wind speed—that is, how large the hurricane is. It also depends on the sea state before the hurricane—whether it was flat and calm, or whether a previous storm had already created waves that the hurricane could augment. Wave heights are not uniform across a hurricane; those on the right-hand side (when the storm is moving away from the observer) will be larger than those on the left-hand side, because they receive an added boost from the forward motion of the storm. Waves will begin to build when the eye of the storm approaches to within 90 to 100 miles.

Once formed, hurricanes generally turn to the north in the northern hemisphere and to the south in the southern hemisphere. This is called *recurving*. The track of tropical cyclones is due to the influence of the deep-layer mean flow in the lower and upper atmosphere. In the North Atlantic, the storms move in response to the large-scale flow around the Azores-Bermuda High in the lower atmosphere and also respond to traveling upper-level disturbances. They initially move westward on the southern side of the high, tending away from the equator. On the western side of the high, a hurricane starts following a northwesterly, then a northerly, track. Continuing to curve to the right, it eventually heads to the northeast. It is possible that at this point its speed will increase, which accounts for the usual path of hurricanes that originate in the Atlantic Ocean near the Cape Verde Islands, North Africa, and track into the Caribbean and the southern United States, as well as those that originate near southern Mexico and track into the Pacific Ocean toward Hawaii, moving around the Pacific High. In the southern hemisphere, after traveling west initially, hurricanes recurve to the southwest and then the southeast. There is no assurance that hurricanes will always follow these paths; there are many exceptions where they have plowed straight west or northwesterly, the Galveston storm in 1900 being an example. The direction of hurricanes is frequently altered by encountering high-altitude winds. Once they reach colder land or water, they lose water and energy and dissipate.

A TYPICAL HURRICANE LIFE CYCLE

The life cycle of a hurricane as viewed from an endangered vessel at sea starts at the bridge, where the captain observes that the barometer has fallen several millibars in the last 20 hours. Weather reports indicate a tropical disturbance moving westward toward the vessel at 10 knots. Shipboard instruments show that the seawater temperature is 29 degrees Celsius (84 degrees Fahrenheit). The sudden drop in atmospheric pressure is significant; when combined with the sea temperature it indicates a tropical storm is possible. The captain contacts the National Hurricane Center by radio and receives the latest satellite photographs.[12]

The storm is now classified as a tropical depression. The satellite photos show the characteristic spiral pattern of clouds, extending out 100 nautical miles with a central pressure down to 990 millibars. Subsequent reports from the National Weather Service indicate maximum winds of 80 knots; the central pressure has dropped to 980 millibars.

The captain makes a sharp turn to the northeast, since his vessel is on the right side of the forward track of the storm. He observes that seas have built to 33 feet. He keeps the wind on his starboard bow and makes as much way as possible at 16 knots, hoping the storm does not

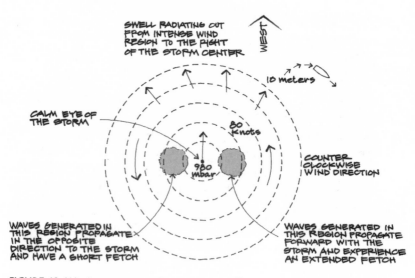

FIGURE 12 Wind-wave patterns from a hurricane.[13]

start recurving. His luck holds; the storm veers slightly to the south-west and he makes port the next day. (See Figure 12.)

By then the hurricane pressure has dropped to 940 millibars, hurricane force winds extend 50 nautical miles in all directions, and seas of up to 40 feet are reported. In two days, gales extend out 200 nautical miles and hurricane force winds out to 75 nautical miles. At this point, the storm has its widest impact; it is recurving north, reaching colder waters and cold air aloft; the winds slowly dissipate and after a few more days the storm dies.

Today, through the use of instrumented sea buoys, airplane flights into hurricanes, satellites, and improved modeling techniques, much more is known about hurricane-generated waves. Measurements typically use an elevation of 33 feet above sea level as a reference point. The data indicate that from this elevation up to, say, 300 feet, the wind speed increases gradually, leveling out at perhaps a 25 percent increase in value. From measurements on various hurricanes, it is known that wind gusts can be around 25 percent greater than the mean wind speed. Thus, at sustained hurricane wind speeds of 64 knots, gusts to 80 knots can be expected. The state of the sea can rise rapidly, even when the eye of the hurricane is as far away as 80 nautical miles. Studies have also been made on correlating the significant wave height with hurricane-strength winds in the open sea. The results indicate that with the onset of hurricane-strength winds, the significant wave height will quickly grow to 26 feet with the potential of reaching 52 feet in a Category Five hurricane with winds of 135 knots. These values can be higher, depending on the sea state prior to the arrival of the hurricane.

There are few actual measured data on hurricane-generated waves in deep water where the sea conditions are violent and most vessels are occupied with remaining afloat and have no time for scientific observations. Now satellites and sensitive transducers on the ocean floor are beginning to provide more data. The available data indicate that hurricane-generated waves can reach extreme wave heights. When Hurricane Ivan approached the coast, it passed over an array of sensors placed on the ocean floor as part of a U.S. Navy research project. At one point waves up to 66 feet high were passing over the sensors every 10 seconds. Winds reached 108 knots. The largest detected wave was 91 feet high.[14] As a hurricane approaches shore, the wave height increases dramatically in shallower water.

So, what are the deadliest hurricanes to have hit the United States? Table 3 lists those recorded during the 104 years from 1900 to 2004.

Hurricane Katrina, which devastated the Gulf Coast region of the United States in August 2005, will be added to this list. It had a low of 902 millibars. The death toll had reached 1,250 and rising as this book was being prepared.

HURRICANE FORECASTING

Much progress has been made in recent years in forecasting hurricane wind patterns. The basic model takes into consideration the forward velocity of the hurricane, the fact that the winds flow inward at some angle (typically around 25 degrees), the central pressure, and the radial distance of the maximum winds (the latter two items being measures of the intensity of the hurricane).[15]

These models yield results that are shown pictorially in Figure 12; the hurricane is moving west (toward the top of the page). This is a

TABLE 3 Deadliest Hurricanes, Continental USA, 1900-2004

Rank	Description	Year	Deaths	Central Pressure (millibars)	Saffir-Simpson Category
1	Galveston, Texas	1900	8,000+	936	4
2	Lake Okeechobee, Florida	1928	2,500+	929	4
3	Florida Keys, south Texas	1919	600+	927	4
4	New England, New York, Rhode Island	1938	600	946	3
5	Labor Day, Florida Keys	1935	408	892 (lowest)	5
6	Audrey, southwest Louisiana, north Texas	1957	390	945	4
7	Great Atlantic, northeast United States	1944	390	947	3
8	Grand Isle, Louisiana	1909	350	—	4
9	New Orleans, Louisiana	1915	275	931	4
10	Galveston, Texas	1915	275	945	4
11	Camille, Mississippi, Louisiana	1969	256	909	5 high
12	Great Miami, Florida, Mississippi, Alabama	1926	243	935	4
13	Diane, northeast United States	1955	184	949	3
14	Unnamed, southeast Florida	1906	164	—	2
15	Unnamed, Mississippi, Alabama, Florida	1906	134	—	3
16	Agnes, northeast United States	1972	122	—	1
17	Hazel, South Carolina, North Carolina	1954	95	938	4
18	Betsy, southeast Florida, southeast Louisiana	1965	75	948	3
19	Carol, northeast United States	1954	60	—	3
20	Floyd, eastern United States	1999	57	—	2
21	Unnamed, southeast Florida, Louisiana, Mississippi	1947	51	940	4
22	Donna, Florida, eastern United States	1960	50	930	4
23	Unnamed, Georgia, South Carolina, North Carolina	1940	50	—	2
24	Carla, north and Central Texas	1961	46	931	4
25	Allison, Texas	2001	41	—	?
26	Unnamed, Texas	1909	41	—	3
27	Unnamed, Texas (Freeport)	1932	40	941	4
28	Unnamed, south Texas	1933	40	—	3
29	Hilda, Louisiana	1964	38	—	3
30	Unnamed, southwest Louisiana	1918	34	—	3
31	Fran, North Carolina	1996	26	954	3

TABLE 3 Continued

Rank	Description	Year	Deaths	Central Pressure (millibars)	Saffir-Simpson Category
32	Unnamed, Louisiana	1926	25	—	3
33	Connie, North Carolina	1955	25	962	3
34	Ivan, northwest Florida, Alabama	2004	25	946	3

Addendum (pre-1900 or not Atlantic or Gulf coast):

1	*Chenier Caminanda,* Louisiana	1893	2,000	948	4
2	Sea Islands, Georgia	1893	1,000-2,000	—	3
3	Unnamed, Georgia, South Carolina	1881	700	—	2
4	San Felipe, Puerto Rico	1928	312	—	4
5	U.S. Virgin Islands, Puerto Rico	1932	225	—	2
6	Donna, St. Thomas Virgin Islands	1960	107	—	4
7	Chubasco, southern California	1939	45	—	TS
8	Eloise, Puerto Rico	1975	44	—	TS

NOTE: TS = Tropical storm.

northern hemisphere storm; winds are rotating counterclockwise (as viewed from above). The winds are highest on the right side of the figure. The waves produced by the storm are a complex combination of both swell and wind-generated seas. Due to the varying direction of the wind, the resulting wave patterns are highly irregular and difficult to model. Waves on the right-hand side of the storm propagate forward with the forward motion of the storm and reach greater heights than waves on the opposite side of the storm. This is because the left-side wind speed is less, the cyclonic winds being reduced by the forward motion of the storm. At the storm's center, the winds are near zero.

A mental image of Figure 12 should be in the mind of every ship captain venturing into tropical waters during hurricane season. For the example cited above, the vessel is shown in the upper right-hand corner of the illustration. A vessel on the upper left-hand side of the illustration would turn to the southwest to escape the storm.

Efforts have been made to make measurements during storms to gather data to validate weather forecasting mathematical models by comparing the model results to actual data. These types of analyses are called *hindcasts* since they represent an "after-the-fact" look at the storm. In other words, forecasters say, "Okay, we measured the wind, waves, and air pressure of a storm; let's put the data in our model and see if the results compare with what actually happened. If not, then how do we improve the model?" As difficult and dangerous as it is to make measurements during major hurricanes, some data have been collected using instruments on offshore oil platforms, islands, and weather ships stationed in the oceans or by using buoys.

The sailing routes from South America to the Caribbean are littered with ancient wrecks of Spanish galleons—ships that sank under the force of giant waves caused by storms while bringing treasures from the New World back to Spain. Many of these vessels disappeared without a trace, leaving no survivors to tell the story. Now we know that given the type of sailing rig they employed, many were literally driven under the water when high winds and large waves arose suddenly before the crew was able to reduce sail. Today, thanks to radio communications and satellite phones, fewer maritime disasters go unreported.

In 1995, Hurricane Roxanne crossed into the Gulf of Mexico, where several hundred offshore oil field workers were on board a large barge that was anchored in place. When seas reaching 30-plus feet began impacting the barge, one by one the anchor cables failed and the barge was set adrift in mountainous seas. An oceangoing tug (normally used to position the huge barge) finally managed to pass a cable to the barge after repeated attempts so it could be taken under tow. The tug kept the barge headed into the violent seas so it would not broach. Before long, the cable parted and the barge continued to be battered by the sea. Massive pieces of equipment—some weighing many tons— broke loose and crashed across the deck like random battering rams. The barge took on water and began to break up, sinking lower and lower in the water. The tug and two other vessels that had risked all to come to its assistance were able to rescue more than 200 crew members who went into the roiling waters as the barge sank. This was truly one of the most miraculous rescues ever made at sea.[16]

STORM SURGE AND HURRICANE-INDUCED FLOODING

An important side effect of hurricanes is the rising sea level, called a *storm surge*, caused by the storm as it approaches shallower water. Although not normally a concern to vessels at sea, storm surge is a serious problem for harbors and coastal installations and can be dangerous to a vessel attempting to make port in a storm.

Storm surges arise as a result of the wind driving water toward the coast and piling it up due to interaction with the nearshore sea bottom and shoreline. Huge storm surges result when a hurricane approaches land with a concave bay and winds are flowing directly into it, especially when the underwater seabed rises rapidly just offshore. The Bay of Bengal has this scenario and is prone to huge storm surges from tropical cyclones that kill so many people.

The low-pressure area in the center of the hurricane adds to the surge height. If we consider that in the eye of the hurricane the pressure is, for example, 900 millibars (equivalent to 674 millimeters of mercury), and remember that mercury is 13.6 times as heavy as water, this pressure is equivalent to a column of water that is 9,166 millimeters, or 9.17 meters, high. Outside the storm, where the air pressure is higher, closer to 1,020 millibars (764 millimeters of mercury), the equivalent height of a column of water exerting the same pressure will be 10.39 meters. The difference between these two numbers (probably a worst case or very near so) is 4 feet (1.22 meters). The pressure effect is relatively minor when compared to the winds piling up water against the shore.

You can picture a hurricane conveying a large bulge of water in its low-pressure center, where the water level could be several feet higher than the surrounding ocean. The height is further increased by the force of the circular winds piling more water into the center of the hurricane. As the storm approaches shallow water near the shore, the storm-driven mass of water piles up even higher. Overall, the wind-driven effect is much more significant than the pressure effect. The water can remain high until the hurricane winds eventually die down.[17]

Hurricanes with high flood waters caused by storm surge include Andrew (1992), Hugo (1989), and Camille (1969), which caused storm surges of 8 to 16.5 feet, and then most recently Katrina (August 29, 2005). This Category Four hurricane caused a storm surge of 26 feet, broke levees, and heavily damaged New Orleans and other Gulf Coast cities. When tropical storms or hurricanes impact low-lying coastal areas, casualties can be extremely high. In fact, it is estimated that 90 percent of the deaths from hurricanes are due to flooding damage and drowning.[18]

Nowhere is this effect more apparent than in the low-lying coastal areas along the Bay of Bengal. Tropical storms and hurricanes have caused a huge loss of life in this region during the past several centuries, for example:

- October 7, 1737, Bay of Bengal, cyclone combined with high tide of 40 feet—300,000 killed
- June 5-11, 1882, Bombay, India—100,000 die
- May 1833, Calcutta—50,000 killed
- October 5, 1844, Bay of Bengal, Calcutta—50,000 killed
- May 28, 1963, East Pakistan, water ran 2 miles inland and carried huge ocean liners 1 mile inland

In September 1900, another hurricane traversed the Gulf of Mexico and roared into the history books by wiping out the city of Galveston, Texas. The storm surge was responsible for much of the damage and most of the fatalities—estimated at 8,000. At its peak, sustained winds were 130 knots with gusts to 170 knots or higher. Waves as high as 40 feet crashed into the city, sweeping away entire houses and multistory buildings. Water 30 feet deep ran through the city streets. Under the tremendous force of the waves, piles of lumber and debris—even entire structures—were pushed inland like giant wrecking balls, destroying all in their path. One man reported narrowly avoiding being crushed by a grand piano hurled at him by a wave. An eyewitness survivor, Dr. Samuel Young, described how the waves entered the second floor of his home, seven blocks from the oceanfront, at a level 30 feet above the street. Moments later the house shuddered and floated free. Dr. Young escaped by using a door as a raft; the rampaging waves carried him clear across the city, twirled him in a whirlpool, and finally wedged the door and him, bruised and bleeding, against a pile of debris.[19]

Today the prudent mariner has a variety of tools—including satellite weather photos and marine weather forecasts—that help avoid sailing into the hazards of storms and massive waves. However, for a vessel already at sea, the options are reduced to steering a course away from a storm or riding it out. To elude the effects of a distant storm, it is necessary to anticipate the likely track of the storm and, most importantly, to know how waves will propagate away from the storm. This is neither an easy nor a pleasant task.

FORECAST ACCURACY

National weather services use several models for predicting the track and intensity of hurricanes. Track forecasts are the latitude and longitude of the storm center, while intensity refers to the maximum sustained surface wind. Forecasts are typically issued for 12, 24, 36, 48, and 72 hours. Two main types of mathematical models are used: one type predicts the storm track; the second type is used to predict its intensity.[20]

Of more interest for our purposes is the accuracy of such models. As might be expected, their accuracy is best in the short-term forecasts and deteriorates for the longer forecast periods. For estimating hurricane tracks, the errors are around 50 nautical miles at 12 hours (with a range of error from 40 to 60 nautical miles, depending on the model). At 24 hours, the average error is around 85 nautical miles, but can be as great as 200 nautical miles in the 72-hour forecast.

Regarding intensity, wind speed errors have recently averaged around 9 knots at 24 hours, 15 knots at 48 hours, and as much as 19 knots at 72 hours. For both intensity and tracks, there are slight differences in the accuracy of the models depending on whether the Atlantic, Pacific, or Indian Ocean is involved. The accuracy of forecasts has improved over the last several decades, but there is still much to be learned about the inner workings of hurricanes.

How accurate are the predictive models in use today? Better than before, but definitely not infallible. Wind is the ultimate determinant of the size and direction of waves. While the ability to predict wind patterns has improved immensely with the advent of satellites, better mathematical models, and radar, it is still far from perfect. Local winds can vary widely from those predicted for large areas. Ultimately, it is the responsibility of the captain to use his experience and judgment to safeguard his vessel and crew.

When the wind speed is already 80 to 90 knots, an error of 10 to 20 percent is not terribly significant; it implies that the resulting waves will be merely terrifying rather than horrendous. However, an error in storm position of 100 to 200 nautical miles could make a huge difference in the strategy of a ship captain desperately trying to avoid an oncoming storm, as was the case of the captain of the *Fantome*.

THE TRAGEDY OF THE *FANTOME*

In October 1998, Hurricane Mitch developed from a tropical storm into one of the largest (Category Five) hurricanes in recent times. Day after day, as the hurricane approached Central America, aircraft from the Weather Reconnaissance Squadron of the U.S. Air Force Reserve Command flew into the eye of the storm and collected data that were relayed to the National Hurricane Center in Florida. These data, plus radar and satellite data, were fed into the center's sophisticated forecasting computers to predict the path of the hurricane and its eventual landfall. But modern weather forecasts are not infallible.

At this time, the sailing vessel *Fantome*—flagship of the Windjammer fleet, a cruise line that employed tall ships—was taking on 90-some passengers for a two-week cruise. Since the first predictions indicated that the hurricane was headed due west toward *Fantome's* base in Honduras, *Fantome* sailed north to Belize City to discharge the passengers at a safe location and to secure the $20 million vessel. Once the passengers were off the vessel, the weather forecasts showed the hurricane turning north toward Belize and the Yucatan Peninsula. The captain and owners of *Fantome* then made the decision to sail the boat south to escape the hurricane. The weather fax onboard *Fantome* was inoperative. Had the captain been able to see the satellite image of the storm bearing down on him, his decisions probably would have been different. As *Fantome* left Belize, the hurricane made an abrupt turn south, almost as if it intended to intercept *Fantome*. The ship turned east, seeking shelter in the lee of the Bay Islands, a group situated in the Gulf of Honduras. Perversely, the hurricane now turned west, aiming directly for *Fantome*, no longer with options for escape.[21]

Up until the moment that *Fantome* foundered and sank with the loss of all 31 crew members on board, the captain was in contact by satellite phone with Windjammer headquarters in Miami. The captain reported that the vessel was taking a terrible beating, rolling and slamming into mountainous seas, waves 30 to 35 feet high breaking over the stern, unable to turn the boat, barely able to keep driving it straight east into confused seas. At this point *Fantome* was around 19 nautical miles from the eye of the hurricane. Aircraft measured wind speeds in the same location at 115 knots, gusting to 150 knots. Such winds would have produced waves with a significant wave height of 46 feet and a maximum of perhaps 66 feet. The terrible irony of this tragic loss is that the hurricane seemed to anticipate *Fantome's* every evasive move, altering its deadly course to intercept the ill-fated vessel. The loss proved one other point: The hurricane also outmaneuvered the best efforts of weather forecasters.

A RECORD HURRICANE SEASON

At the start of the 2004 hurricane season, the National Weather Service issued its outlook for the Atlantic-Caribbean region, stating that "above-average" activity was to be expected. In retrospect, this was an understatement of what actually happened, as 2004 turned out to be extraordinary in that four hurricanes hit the southeastern United States in quick succession.

First came Hurricane Alex on the South Carolina coast in July, followed by tropical storm Bonnie in August. As Bonnie moved toward Florida, tropical storm Charley arose and also began moving toward

Florida, where it soon reached hurricane strength. Charley was merely a preview of things to come.

Subsequently, during the months of August and September 2004, the Caribbean area, Florida, and the southeastern United States suffered from the impact of several more large hurricanes (Frances, Ivan, and Jeanne). Charley caused $14 billion in damage and killed 10 people. Indirectly, another 20 U.S. deaths were attributed to Charley, along with 5 people killed in the Caribbean. Not since Hurricane Andrew in August 1992 had such damage occurred, and never before had four major hurricanes struck in close succession. The hurricanes arrived one after the other, one to two weeks apart. Barely had residents begun to dig out from under the damage from one hurricane when they were besieged by another. In addition to experiencing winds from 78 to 130 knots, coastal areas were impacted by extreme waves and storm tides that destroyed waterfront installations, tossed boats ashore as if they were so many toys, and altered entire sections of beach and shoreline. Total damage was in excess of $40 billion. These four hurricanes were in the range of Category Two to Four when they hit the United States, although Ivan reached Category Five several times in the Caribbean. Ivan was responsible for 25 U.S. fatalities. The 2004 season was terrible, but 2005 was worse. Hurricane Katrina (Louisiana and Mississippi, August 29, 2005), Category Four, is now estimated to be the worst natural disaster in the history of the United States.

In addition to their impacts on the United States, the hurricanes caused widespread damage throughout the Caribbean. For example, Hurricane Ivan resulted in considerable destruction on Grenada, hitting the island with 100-knot winds and then moving on to become a Category Five hurricane. Ivan also damaged Trinidad, Tobago, Jamaica, and Grand Cayman, before recurving to hit Alabama and the Florida Panhandle.

In terms of size, longevity, and total destructive power, the hurricanes listed in Table 4 stand out as some of the worst experienced in recent times.

If hurricanes represent the worst case of storms causing large waves, how big can waves get? In all likelihood, those who could have answered this question perished in the storm. Based on the available measurements and data, a height of 90 feet certainly seems plausible.

PLATE 1 *Dan Moore surfs a 68-foot-high wave.*

PLATE 2 *Tsunami damage at Hilo, Hawaii, May 22, 1960.*

PLATE 3 *Bobsled* in the 1998 Sydney-Hobart race.

PLATE 4 *Crew prepare to abandon dismasted and sinking* Stand Aside .

PLATE 5 (A) A brisk sail; (B) Wind-blown wave in a storm.

PLATE 6 Tidal bore, Qiantang River, China.

PLATE 7 The wreck of the Memphis in Santo Domingo (1916).

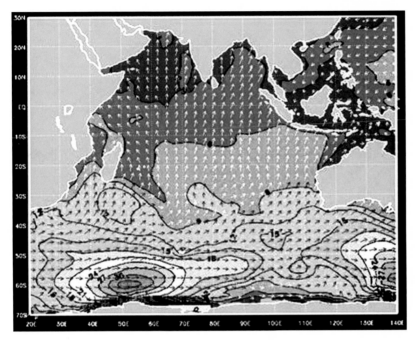

PLATE 8 U.S. Navy Indian Ocean wave forecast.

PLATE 9 Tsunami flooding at Laie Point, Hawaii, March 9, 1957.

PLATE 10 Man caught in tsunami, Phuket, Thailand ,December 26, 2004.

PLATE 11 Tsunami-damaged bungalow, Koh Phi Phi Island, December 26, 2005.

PLATE 12 *Wave victim on the Agulhas Coast, South Africa.*

PLATE 13 *Freighter in a typhoon, North Pacific.*

PLATE 14 Merchant vessel Winter Water *fights heavy seas from tropical storm* Georgette, *Eastern Pacific Ocean, July 28, 1980.*

PLATE 15 *North Pacific Ocean storm waves as seen from M/V* Nobel Star, *winter, 1989.*

TABLE 4 Worst Hurricanes

Name	Year	S-S Category	Maximum Winds (knots)	Damage (billion dollars)
Katrina[a]	2005	4	150	100-200
Andrew	1992	5	125	44.9
Betsy	1965	3	118	6.5
Camille	1969	5	139	14.9
Gilbert	1988	4	101	3.0 (approx.)[b]
Hugo	1989	4	140	12.7
Iniki	1992	4	125	1.8
Luis	1995	4	120	2.5+[c]
Mitch	1989	5	155	Unknown[d]
Opal	1995	5	130	4.1

[a]Category Four at landfall; Category Five in the Gulf of Mexico.
[b]Lowest recorded pressure, 888 millibars.
[c]*Queen Elizabeth 2* was on the fringes of this storm when hit by a rogue wave.
[d]9,086 dead, mostly in Central America, including crew of 31 from *Fantome*.

Wave heights during storms are affected by the condition of the sea before the storm. Sometimes called *sea severity*, the sea condition influences how big the waves can become. For example, if winds have been blowing before the storm, waves will be larger than if the storm originated in calm seas. Ochi cites two cases in which winds of 12 to 27 knots had been blowing for up to 10 days prior to a storm. Waves were already 8 to 16 feet high. With the advent of the storm, they increased to 49 to 56 feet in a matter of 21 hours.[22] Thus, in a fully developed sea, a smaller hurricane might produce larger waves than otherwise would be expected.

The National Weather Service anticipates that hurricane activity in the Atlantic-Caribbean area will continue to increase during the next several decades. Naturally, this prediction is of grave concern to mariners who out of necessity find themselves in these waters during the hurricane seasons.

There are others, however, who monitor the ocean's great storms from afar and eagerly await the arrival of the resulting waves on distant shores. Rather than run from huge waves, they run after them, hoping to ride them on flimsy fiberglass surfboards.

5

Swell

If you drive along the California coast, you will often see the backs of figures in wet suits seated on surfboards, looking out to sea for the "perfect wave." This same scene can be observed on any given day on coastlines throughout the world. What is it that brings these athletes to the fringes of the world's great oceans, often early in the morning, even on days when the weather is miserable? It used to be that someone in the beachfront community passed the word—surf's up—and chores or jobs were set aside in a race to get into the water and claim a good spot to catch a wave. Today, "surf's up" is more likely to come from the Internet, where dedicated surfers can use their web browsers to navigate to a web site that provides not only the latest marine weather forecasts—including expected wave heights and periods—but also real-time data from ocean buoys thousands of miles away and even from video cameras that scan favorite beaches so a surfer can see what the waves look like without first driving to the beach.

Surfing information on Internet sites helps surfers keep track of distant storms, use ocean buoy data to track the height and movement of waves, and use wave propagation models to predict when waves from a recent storm in the North Pacific will finally reach shore in Newport Beach, California; Hawaii's North Shore; Mexico; or Tahiti as the long, undulating waves known as *swell*.

SWELL

Just what is swell? As I explain in Chapter 4, storm winds typically follow a circular path, stirring up the sea as the storm propagates. Storm waves mix with preexisting sea waves, and in a short time the sea roils with a mixture of waves large and small, with long and short wavelengths, coming from various directions. This scenario is known as a *confused sea*. (See Chapter 7 for more on this subject.) Waves radiate from the storm center, in much the same manner as they emanate from a rock tossed into a quiet pond. Some of the waves will head off in the direction in which the storm is moving, in which case, if the storm continues to blow, they will grow larger. Others, headed in the opposite direction, will not increase in size. In either case, storm-generated waves propagate from the center of the storm, the faster waves (those with longer periods and wavelengths) gradually outstripping their slower counterparts. In this process, known as *dispersion*, the sea acts as a filter so that at locations distant from the source, the waves that traverse a particular point eventually will exhibit similar characteristics. These distant, dispersed waves are known as *swell*. (Later, when the slower waves reach this same point, their characteristics will be different.) Swell travels long distances, only slowly giving up its energy. Swell from a given North Pacific storm can be tracked clear across the Pacific until it finally dissipates on the coast of North America.

Often the yachtsman will note that swell does not consist of nice uniform waves, but instead has a set of waves that seems to come at one angle to his course and another set that comes at a different angle. When swells from two different sources interact, the two (or more) wave fronts can produce various patterns. For example, two swell wave fronts intersecting at 45 degrees would produce a diamond-shaped pattern of crests and troughs. There can be more than two swells interacting—for example, near islands where swell impinges on the shore and reflects or is refracted back into the sea. If two sets of swell meet at right angles, a checkerboard pattern is produced by interference of the individual wave trains. It is not unusual to encounter wave conditions that are a mixture of swell with superimposed wind waves. In these cases there is no standard pattern to wave crests and troughs, and this also can be called a "confused sea."

Swell eventually sorts itself out into one regular pattern of waves, trough following crest in an endless roll across the open ocean until land is reached. In this process, swell loses very little energy until shallower water is reached and the waves begin breaking. As it crosses the ocean, if it encounters favorable winds, swell can grow in height.

I can recall standing on the beach on idyllic summer days and looking out to sea at the seemingly endless progression of waves rolling in, swell generated by some distant Pacific storm. Surfers call this a *corduroy sea*, the term inspired by the distant blue-green ridges of water. When body surfing—that is, without a board—I looked for waves big enough to give me a good ride but not so big that it was impossible to swim out to them or so big that they would grind me into the sand if I caught them at the wrong time.

LARGE SWELLS

Historically, certain years stand out as years of giant swells in the Pacific—1953, 1969, 1983, and 2001, for example.[1] Of these, the swell of 1969 is perhaps the most notable. In late November 1969, a large storm started building in the North Pacific Ocean, near the Kamchatka Peninsula. Moving east, it combined with another storm and increased in size to the point at which winds greater than 50 knots were blowing along a 1,700-nautical-mile front reaching from north of Hawaii to the Aleutian Islands. Then it stopped and remained stationary for a day and was joined by a third storm. With high winds, lengthy duration, and a fetch of thousands of miles, it is no surprise that huge waves were produced.

On the first of December, 30-foot-high waves hit the northern island of Kauai; hours later, Oahu's North Shore was hit by waves up to 50 feet in height. Some areas were evacuated, homes were destroyed or badly damaged, coastal highways were flooded, and several people were caught in waves and drowned. Four days later the big waves reached Southern California, where there was a similar pattern of beachfront damage, flooding, and several deaths. In the midst of all of this chaos, a few highly skilled surfers rode waves in the 18- to 35-foot class, among the highest ever ridden up until that time. Most beachgoers just stood on the shore and stared at the huge waves as they came crashing ashore.

Another big swell came in 1983. More than 1,000 structures along the Southern California coast, including piers in Oceanside, Seal Beach, Redondo Beach, and Santa Monica were damaged by the waves; about one-third of the Santa Monica pier was lost to the sea. It was the swell of 1983 that saw the inauguration of big wave surfing.

SURFING

Board surfers live for those magical days when a large swell builds along the coast from distant storms and the perfect ride seems attainable. Those who live close to the beach can walk or drive to their favorite spot and look at the line of waves rolling in to decide if it is worth it to unlimber their boards and put on their wet suits. Others, like my son Kent Smith, a biochemical research scientist for Mannkind Corporation, access web sites such as *www.surfline.com*, where they can see the latest projections of surf conditions, even being able to see the waves in real-time cameras mounted at the most popular sites. It is also possible to access weather buoys located along the coast and in the ocean, to check wave heights and periods, and to track the likely progression of different storms. Is the swell building or dying? A good storm might create conditions that last several days. When the waves first reach the coast, they will be building and sets will come closer together as the day progresses. Later, once the effects of the storm begin to dissipate, the wave energy starts decreasing and sets are spaced farther and farther apart, leaving the surfers to sit motionless on their boards, scanning the horizon for a sign that another wave is coming.

Surfers have their own lexicon for describing wave heights and extreme waves, comparing them to their height when standing on their boards. A waist-high wave is hardly worth getting suited up for, unless the beach is uncrowded and other conditions are perfect. Head-high waves draw surfers like a magnet to the best spots; when the waves are said to be "overhead," the less hardy begin to drop out. "Double-overhead" waves represent a cutoff point; a 20-year veteran like Kent would surf if conditions were perfect, but would exercise caution on a stormy day when the seas were uncertain. When you read of the surfers in Hawaii who ride 25- to 35-foot waves, you begin to get a real appreciation for the skill and risk that are involved. And a 68-foot wave, as described later in this chapter, is something else, totally out of the league, akin to climbing Mount Everest without oxygen.[2]

It is not just the height of waves that matters as they come rolling in from storms in the distant reaches of the Pacific, but how they impinge on the shore. Surfers describe beach breaks, point breaks, or reef breaks. A point break is where the incoming swell is refracted by a point of land, a cape, or even a man-made jetty. In Southern California, locations where point breaks occur have descriptive names: "Trestles," "Rincon," or the "Wedge."

When conditions are right, point breaks give long, exhilarating rides. By catching the wave as it starts to break near the point, the surfer is able to ride the cresting wave in a direction parallel to the beach for a long distance. When the surf is high, one of the challenges is getting to it. With a point break, surfers can usually paddle their boards around the outside of the break and catch the next wave. At a beach break—as at Huntington Beach or Black's Beach near San Diego—surfers ride the waves into shore and then paddle back out through the surf to catch the next wave. This becomes a challenge with large waves—you somehow have to fight your way through them to get back out for the next ride or paddle out around the break. Sometimes the outgoing current is channeled in such a way that it facilitates the ride back out.

Wind plays into the equation as well. Surfing in a storm—in chaotic seas stirred up by the wind—is not enjoyable. Neither is a strong onshore wind, because it tends to flatten the waves and reduce their height. In Southern California there are many days when the land heats up, there is a high-pressure area inland, and winds blow offshore, a condition locally referred to as *Santa Ana winds*. This is highly desirable, not only because it usually heralds a warm day at the beach but also because the wind tends to blow against the front face of the incoming waves and hold them up so that they don't break, or take longer to break.

When a large wave breaks over a shallow area, such as a reef or sandbar, the bottom of the wave is slowed and momentum carries the top of the breaking wave forward so that it curls over and creates a hollow tunnel, called a *tube* in surfing jargon. A strong Santa Ana wind also helps form a tube because the wind lifts and supports the crest as it starts to fall. Good, rideable tubes occur infrequently, but when they do, a spectacular image is created when the surfer disappears behind a huge waterfall of seawater and emerges unscathed a few seconds later at the other end—surfing nirvana, you could say.

Reef breaks as in Hawaii and other island chains present opportunities for really large waves—say, double or triple overhead—but conditions must be considered carefully. First, the tide must be just right. Too low, and the waves might break on the reef itself, presenting the potential for serious injury to the careless surfer. Too high, and waves pass over the reef without breaking. When conditions are right, waves coming in from deep water pile up as they reach the reef, creating the long and impressive tube rides seen in surfing movies during which the rider seems to be traversing a long tube of water, the sea towering over his or her head.

How about a simulated reef break in midocean? There are instances in which a rogue wave has suddenly risen up out of the water and struck a ship broadside. The result is most likely a capsize, unless the vessel manages to recover from the roll. If a wave rose up in front of a vessel and hit the bridge, the effect would be quite different. The force of the water is so great that it can shatter the thick glass of the pilothouse windows, and there have been cases in which the wave washed the helmsman off the bridge or tore the bridge off the vessel. In other cases, plowing bow-first into a huge wave, vessels weighing thousands of tons come to a nearly complete stop, the impact causing the steel vessels to shudder over their entire length. I cannot imagine anything more terrifying than being on the bridge of a large vessel at night when something like this happens. Yet there are other brave souls who deliberately confront such giant waves on surfboards.

EXTREME SURFING

These daredevils are members of a very select group of world-class surfers who use personal watercraft—wave runners—to haul surfers out to catch a giant wave and then recover the surfer after the ride. When the waves are extreme, they are moving so fast that it becomes impossible to catch them by paddling into them by hand on a surfboard. Instead, surfers are towed into the wave and released. I thought it would be useful to talk to some of the best of these extreme wave surfers, and see what I could learn about the behavior of giant waves from people who have experienced them up close—who have an intimate knowledge of them.

Cortes Banks, described earlier, is in the category of a reef break. When winter storms create 50- to 100-foot-high waves, surfers must first make a 100-nautical-mile boat ride just to get to the site. Once at the site, they are towed into position by wave runners racing at speeds of around 40 miles per hour. After the ride, the wave runner circles back, retrieves the surfer, and takes him or her out of harm's way—assuming he or she survived the ride. Mike Parsons is a Californian who surfed the big winter swell in 2001 at Cortes Bank, riding a wave estimated to be 60 feet high.[3]

There are other spots where surfers go to seek big waves—in Northern California, Australia, South America, and of course, the North Shore of Oahu, Hawaii. In February 1986, several surfers rented a beach house on the north shore at Ke Iki. The swells were running 15 to 25 feet with occasional waves to 35 feet. At midnight on February 22, a giant wave hit the beach, threw cars across Kamehameha Highway, and destroyed four houses. Bruce Jenkins and Rich Stevens were sleeping in one of the houses; Rich literally surfed out of the house on a mattress when the wave blew the house apart, ending up in a parking lot 50 feet away. Bruce was trapped momentarily under his bed, but managed to work himself free. Later, eyewitnesses estimated the size of the wave at around 60 feet.[4]

When the last El Niño hit, winter swells were massive in Hawaii. On January 28, 1998, surfer Ken Bradshaw was towed out into the huge surf at a spot called Log Cabins (outside the reef at Waimea Bay), where he successfully caught, rode, and survived a wave that was estimated to be 80 feet high. Dan Moore—also a surfer—was there to see it, and I asked Dan if he thought that it was the largest wave anyone had ever ridden.

"Ken and I were partners," Dan said. "We surfed together for a number of years. In fact, he was the one who got me interested in towsurfing. That day at Log Cabins was incredible. They were some of the biggest waves I'd ever seen. Ken caught one that was just unbelievable. As for anybody riding something bigger than that, I don't know, maybe at Jaws (Pe'ahi, Maui) on January 10, 2004, when Laird Hamilton caught a monster wave; it could have been bigger."

This is a dangerous sport; the celebrated surfer Mark Foo drowned in Northern California in surf with smaller waves. Mark was a consummate, experienced surfer, renowned for his unique and graceful style. His tragic death is a reminder of just how vicious the sea can be, even to the most skilled surfers. Others have suffered broken backs, broken necks, and other injuries after being driven into the sandy bottom or, worse yet, a coral reef.

TRACKING BIG SWELLS

Figure 9 in Chapter 3 is familiar to all boaters—a National Weather Service wind-wave forecast for March 7, 2005.[5] Note the storm building in the Northeast Pacific with 42-foot-high waves centered at 150 degrees west longitude, 40 degrees north latitude. By coincidence I happened to save this particular forecast as I was writing this chapter. Several days later, Kent Smith, whom I'd consulted on technical questions regarding surfing, called to say that a large swell was impacting the coast at Ventura, California, where he lives. Waves were expected to reach 13 to 16 feet. The *Los Angeles Times* published an article describing big waves and the possibility of flooding in low-lying coastal areas.[6] The storm had initially been around 1,500 nautical miles from Ventura. When I looked at the wind-wave forecast for March 10, the storm had moved northeast and was close to Valdez, in the Gulf of Alaska. The distance was now about 1,700 nautical miles. Thus, the swells originating in this storm took 72 hours, traveling at 24 knots, to reach the beaches at Ventura.

There are a number of sources for wave information. One of the best is a U.S. Navy web site, commonly referred to as the "WAM" site.[7] Plate 8 shows a WAM forecast for April 16, 2005, at 0 hours "Zulu" (Greenwich) time. The biggest waves—36 feet—can be seen at latitude 60 degrees south, off the coast of Antarctica, propagating north past Madagascar, in the opposite direction of the Agulhas Current, but no big waves are indicated in this region in the forecast.

The National Oceanic and Atmospheric Administration, as well as Canada, the Scripps Institute of Oceanography, and other entities operate a series of buoys in the Northwest Pacific and California coastal areas. These also can be accessed using the Internet. I did this for buoy 46006, which bears the name of "SE PAPA." It is a fixed buoy located 600 nautical miles west of Eureka, California. I could see a clear pattern of the wind speed building to 29 knots on March 7, 2005, with gusts to 35 knots, a significant wave height of 16.4 feet, and a period of 14.3 seconds. On March 10, 2005, the buoy at Goleta Point, California, was showing the predominant swell arriving from the west (around 260 degrees), with a dominant period of 18 seconds and a significant wave height of 9 feet. Here again, the height was steadily building. Farther south, near La Jolla, California, a site predicted rising surf on March 9 and 10, possibly 10- to 14-foot wave faces. For March 10, the forecast concluded with a final note: *Proceed with caution.*

RIDING EXTREME WAVES

Farther north, at Monterey, California, near the famed Pebble Beach golf course, "proceed with caution" was the watchword for the day on March 9 at a little-known surf break called "Ghost Tree," at Pescadero Point. Here, surfing contestants were competing in an event called the annual Billabong™ XXL Global Big Wave Awards. The idea of the contest is to search around the world for the biggest wave you can find, ride it without killing yourself, get someone to take a picture to prove you did it, and get paid $1,000 per foot of wave height if you survive and are a winner. Among the finalists, several had already ridden big waves at "Jaws," a spot for big waves at Pe'ahi on the island of Maui, Hawaii. Previous winners have ridden waves at Mavericks, located near

Half Moon Bay, California, or have chased giant winter swells at Cortes Bank.

The March 7 storm described above was cooperating nicely with the competition. Monterey was about 165 nautical miles closer to the storm and faced directly into the waves, not sheltered behind Point Conception as was Ventura. By March 9, waves with faces of 60 feet or more were crashing on the shoreline with such violence that golfers on the nearby golf course stopped to watch. By the end of the day, the score was Ghost Tree 2, surfers 2. One surfer sustained a shoulder injury, and he and his rescuer had to be rescued; another broke his leg in four places. Two men managed to ride the largest waves and made it to the group of five prize money finalists. However, the winner turned out to be Dan Moore, credited with riding a 68-foot-high wave at Jaws (Pe'ahi, Maui). (See Plate 1.) The previous winner was Pete Cabrinha, who received $70,000 for his world record ride on a 70-foot-high wave in 2004.

When I spoke with Dan, I asked him how it felt to ride a fiberglass board 6 feet long weighing about 10 pounds in the presence of huge waves that were known to break the backs of supertankers and container ships built of steel and weighing thousands of tons. He described it as exhilarating, an adrenaline rush.

"It is a remarkable experience, just entering the wave; it is an entirely new environment that nothing can quite prepare you for—the wind blowing, hurling salt spray in your face, partially blinded, being hammered by jets of water. Just negotiating the chop to get into position you take a beating, a lot of shock absorption. It looks easy, but the average person, without training and conditioning, could not deal with this. Then the release, and you are racing down the face. It looks smooth, but it is actually a bumpy, jarring ride, until you get to the trough; there it smooths out.

"While it seems longer, the entire ride is over in about 45 seconds.

"You really have to rely on your partner to get you into the wave and then get you out. Mark Anderson drove for me at Jaws. He has years of judgment, of being on the water watching waves, and is able to anticipate what the wave will do. You have to know where the wave is going, know how the wave runner is traveling at 20 to 30 miles per hour, where the wave will be when the wave runner comes from be-hind to catch it, and finally you have to correctly judge the moment of encounter to put the surfer on the tow rope exactly at the 'sweet spot.' Timing is critical at the point of release. There is kind of a whipping action. The tow rope slings you into the wave; you accelerate up to 35 miles per hour. You have to enter the wave just as it starts to get steep—not too early, not too late—at a high enough position to be able to run down the face as it curls over behind you and yet still be able to get out of the way before it buries you.

"Your partner hangs back behind the wave, looks for you to come out, or if you crash, looks for you to pop up, then races in to get you on the sled and get out of there before the next wave breaks. On this wave, my board didn't release properly and I ended up fracturing my ankle. I got hit by another wave and then somebody came in and got me out.

"Timing is important with the wave runner also—blow it, get caught by a wave, and you can kiss it goodbye. Last season I think there were five that got creamed, ended up on the rocks. That's $10,000 each. Then, to add insult to injury, you have to hire a helicopter at $700 to come pick it off the rocks. No littering the beach allowed!"

I asked Dan about his worst wipeout.

"Worst one—which time? As they say, you have to break a few eggs to make an omelet. I've had a lot of spills. That's probably the only way you can learn. But I'd say on January 19, 2004, at Jaws, my first wave of the day was the most recent 'memorable' one. I went down in a bad way. I was on a new board; the wave was 20 to 25 feet high. I went into it and something happened to the board—maybe cavitation. I did an immediate face plant, then got sucked up over the falls. I was annihi-lated. It is hard to describe the violent motion that goes on inside a big wave. You have tons and tons of water pounding on you. When water hits at high speed, it is hard, very hard. It drives you down into caverns at the bottom of the ocean and then forces you back up under pres-sure. I don't think I hit anything. It was just that the forces cause your brain to rattle around inside your skull. When I finally came up I was disoriented, seeing stars, dopey. I must have had a mild concussion. Two more waves dumped on me, until finally someone came in and got me out. I don't remember much; I just saw a blur of motion, somebody on a wave rider who yelled 'raise your hand.' I was so out of it I didn't even know to raise my arm. He grabbed me and pulled me onto the sled. He took me to one of the photo boats nearby where I rested. It took me about an hour and half to recover. Then I grabbed my backup board and went back in and got another dozen rides or so, and one more wipeout.

"This was just one of a number of memorable experiences, all part of the learning process."

TOWSURFING

Dan and many of the other extreme wave surfers belong to the non-profit Association of Professional Towsurfers. I spoke with Eric Akiskalian, a cofounder and current president of the association, about what it took to excel in this dangerous and demanding sport. He told me how surfing emerged as a new sport when surfers began pursuing ever-larger waves, using such personal watercraft as the Yamaha High Output that has a 160-horsepower engine, can carry one to three riders, and can travel at speeds up to 60 miles per hour. He remarked that the main qualities are training and physical conditioning, adding that the ability to read the waves and anticipate their behavior is very important, but always expect the unexpected.

Akiskalian grew up in Santa Barbara, California, and has been surfing abroad for more than 35 years. (In addition to cofounding the towsurfers association, he created an extreme surf web site, *www.towsurfer.com.*) I asked Eric to give me some background on towsurfing and how it got started.

"With the inherent dangers involved and the ever-growing interest in extreme sports, tow-in surfing has become one of the most exciting competitive water sports in the world. It didn't happen overnight; it began as an experiment by Laird Hamilton and some other Hawaiian surfers, including Buzzy Kerbox and Derrick Doerner. They were the first to take an inflatable Zodiac raft with a 40-horsepower motor out to a spot called Phantom's on the north shore of Oahu in the early 1990s."

Their approach was simply to check it out, according to Eric. "They didn't even tow that day. They just went out in the boat, looked around, while dropping into a 15-footer that almost ran them over. It was a little creepy, and if they'd been caught and flipped with the engine blazing, it could have been nasty."

Eric told me that it was not until 1991, when Buzzy and Derrick made an attempt with a 60-horsepower Mercury Outboard, that they actually started to get the hang of a motor-assisted tow, like a water-skier or wakeboarder. It gave them a quick enough start so that they had the speed to glide down the giant open face of the wave after release. They got the idea of using a personal watercraft, or PWC, from these early experiments. The following year they brought their experiences and passion to Pe'ahi on Maui's north shore (the spot called Jaws) to tow into surf even bigger and more powerful.

"As it has evolved, tow-in surfing is an ocean-based sport that requires the use of a PWC, rescue sled, life vests, tow rope with handle, and two very experienced and passionate big wave surfers," said Eric. "Drivers use the PWC, trailing a 30- to 40-foot-long rope and handle, to position their surfer in the right part of fast-moving ocean swells. When the surfer drops the rope, he uses his momentum to catch waves that can't be caught by paddling.

"Drivers monitor the position of their surfer at all times and place the PWC just behind the breaking wave to offer immediate assistance to a surfer who finishes or falls on each wave. Buoyant rescue sleds are attached to the back of each PWC, providing a stable platform for surfers to grab hold of and travel in and out of the wave lineup.

"The surfer's life depends on his partner's ability to drive that PWC, assist in pickups, and come in for the intense rescue before the next mountain of water rolls over them. It is not uncommon for surfer, driver, and PWC to get plowed over by four or five building-size walls of whitewater. I should emphasize that tow-in surfing is a dangerous sport that requires a level of mental and physical conditioning that only year-round conditioning and training can provide. There has to be a bond and a high degree of confidence between the partners. Mental and physical preparation can include underwater rock training (holding or carrying a rock underwater to practice breath holding), swimming, weight training, and paddling, paddling, paddling—in short, everything to insure superb physical conditioning."

So where are the top big wave spots in the world, I asked? Eric mentioned a handful of locations where big waves occur:

- Maui, Pe'ahi—Jaws
- Northern California—Mavericks
- Mexico—Todos Santos
- Southern California outer waters—Cortes Banks
- California central coast—Channel Islands
- Northern California, Monterey—Ghost Trees
- Tahiti—Teahupoo
- Oregon—Nelscott Reef
- Australia—Shipsterns Bluff
- France—Belherra Reef
- Chile
- Canada
- Africa—Dungeons

I also asked Eric if there is an ultimate limit in the sport—if he thought that surfing a 100-foot-high wave is possible.

"Of course," he said. "Someone will do it. It is just a matter of finding the wave, being in the right place at the right time. If they can't get towed in, they'll drop from a helicopter or find another way, but someone will do it. Just like Mount Everest—once you knew it was there, someone had to climb it, even though you might get killed in the process."

LONG-PERIOD TRAVELERS

The ability of swell to travel long distances was demonstrated dramatically to me in September 2005. Around September 8 or 9 an extratropical storm developed near New Zealand. I heard of this from my friend Ray Holdsworth, who had just returned from a cruise in Tahiti. During a stop in Bora Bora, he learned that on the night of Saturday, September 10, a large wave or waves had washed over the reef and damaged some resort buildings, resulting in an evacuation of guests in the early morning hours. A few days later the swell reached Hawaii and created unusual surf conditions. Meanwhile the California newspaper surf reports were predicting that a big southerly swell would arrive eight days after the New Zealand storm on Friday and Saturday, September 16 and 17. Thursday, September 15, had a 6.5-foot-high tide forecasted at 8:00 P.M. for the Southern California coastal area, with the tide being slightly higher Friday night when the moon was at its perigee (closest approach to earth). Ironically, the arrival of a big swell coincided with one of the year's highest high tides.

At that same time, Bill Watkins, vice commodore of the California Yacht Club, along with other members of the club, had his boat on a mooring in Catalina Harbor. He told me that on Thursday night a large wave came into the harbor, rocked all the boats, and tossed dishes and other supplies onto the floor of one of them. This in itself was unusual, since Catalina Harbor is a deep protected cove where one rarely feels any boat movement at anchor or on a mooring. It was made more unusual by the fact that the wave also washed out a portion of the road leading to the California Yacht Club's Ballast Point facility at the edge of the harbor or, as Watkins put it: "Just like you'd used a skiploader to cut a channel." Around the same time Thursday evening, large waves crashing on shore broke windows in oceanfront homes at Malibu, California, and on Friday and Saturday surfers had 12- to 15-foot-high waves at Newport and other south-facing beaches in Southern California. Remarkable to think that waves generated by a storm in the southern hemisphere could retain their energy and travel 6,000 nautical miles at an average speed of around 31 knots and cause damage in the northern hemisphere on the opposite side of the Pacific Ocean!

SWELLS IN MIDOCEAN

One of the joys of being on the ocean on a good sailing day is sailing "down wind and down swell." Under these conditions the boat moves smoothly, a slight heel, and the helmsman can feel a surge of speed as each swell passes under the stern of the boat and pushes it toward its destination, adding to the forward motion due to the wind. Looking aft, the helmsman can see corduroy—an endless progression of waves coming from the horizon toward the vessel.

Normally, because of its long wavelength, swell does not represent a hazard to oceangoing vessels. Even if the significant height is large, the vessel will ride up the wave front and slide gradually down the back side. However, if waves having a similar wavelength intersect, some crests can add, while others cancel out, and larger waves can be produced. This condition can give rise to extreme waves and is discussed in Chapter 8. Sometimes such large waves will break in midocean, creating new, shorter-wavelength waves. The helmsman, lulled into complacency by the endless succession of uneventful swell, will suddenly experience more dramatic action as the ship is tossed about by the shorter-period waves. A common remark muttered under these circumstances (as the galley crew picks the dishes up off the floor) is, "Where did that come from?"

Today, satellite observations of the oceans are leading to improved understanding of swell patterns throughout the year. By using satellite-based altimeters to measure wave height and by making simultaneous satellite determinations of wind speed, global swell probability maps can be constructed. These maps indicate a northward trend in ocean swell patterns in the Pacific and Atlantic oceans. This is thought to be due to the northward propagation of strong swells produced by winter storms in the southern hemisphere during the austral winter. Meanwhile, the swells in the Indian Ocean tend to extend southward in spring and westward in winter, but diminish in area during the summer.[8]

Tod and Linda White, friends and neighbors of mine, sailed their 37-foot sailboat *Seascape* to Hawaii during the early summer, following a long, curving, westerly arc from 33.5 degrees to 20.5 degrees north latitude. Encountering nearly perfect weather, they made the 2,200-nautical-mile trip in just 16 days. Tod reminisced about standing watch one night when a 10-foot swell was running.

"It was dark—just stars and a new moon, but enough light to make out the waves as they approached *Seascape*'s stern. I'd see a dark wall of water rise up higher than my head, if I'd been standing. It would block my view of the ocean behind us. Magically, the boat would rise up as the wave swept under *Seascape* and passed us on its way, with nothing to note its passing other than the feeling of *Seascape* sliding down the rear of the wave to its former position, and a soft swishing sound as the wave flowed past the boat. Behind us, the sea glowed faintly in a trail of phosphorescent bubbles and then all was still until the next swell appeared. There was a special peacefulness to it, alone in the Pacific, far from land, the only sound being the boat's movement through the water and the sensation of mile after mile sliding under the hull, wind and wave bringing us ever closer to landfall at Hana, Maui."

The long, smooth undulations of the sea are familiar to every sailor. Thus, on August 27, 1883, the captain of the vessel *Evelina* was not particularly concerned when he observed some large, smooth oscillations of the sea at his location near the Cargados Carajos Shoals, a remote speck of ocean reefs in the Indian Ocean several hundred nautical miles northeast of Mauritius, at latitude 17 degrees south and longitude 60 degrees east.[9] The only thing surprising was that the sea beforehand had been calm, and this disturbance occurred shortly after noon. If the captain had been able to send a diver down, he would have noticed another curious fact: The water deep below the ship was moving, rather than being calm as it would have been in the case of a gentle surface swell. Later he would learn that 2,660 miles to the northeast, the island of Krakatoa had just blown itself to bits. Traveling at 320 knots, the tsunami wave had just passed under his ship.

6

Terror Waves: Tsunami

The ancient Polynesians believed that earthquakes were caused by the god Ruau-Moko, the youngest son of their ancient gods. Ruau-Moko had remained in the Earth Mother's womb and it was his movements that caused earthquakes. Other ancient civilizations had similar legends; some believed the earth is supported on the backs of turtles or, as in Mongolia, on the back of a giant frog. How else could we explain how solid earth and rock could suddenly move?

THE RING OF FIRE

Today we know that earthquakes result from the movement of tectonic plates—massive layers of stone deep within the earth that over the ages slowly move and grind against each other. At some point, the enormous forces overcome the friction of rock against rock, and one plate will suddenly move relative to another, deep within the earth, or under the sea, where a tsunami can be produced. The Pacific Ocean is bordered on all sides by the intersection of such plates, giving rise to the numerous earthquakes that occur in Chile, California, Alaska, Japan, and other locations on the edge of this great ocean. Because of its fre-

127

quent seismic and volcanic activity, this zone that circles the Pacific is known as the "Ring of Fire." On the east, the Pacific Plate encounters the Nazca Plate in the south and the North American Plate in the north. On the west coast of South America, the Nazca and South American plates come together and are the source of considerable seismic activity there. In the southwestern Pacific, it is the intersection between the Pacific and Australian plates that gives rise to seismic activity. The strong earthquake belt that runs roughly east-west through the Mediterranean and Central Asia occurs along the intersection of the African and Eurasian plates.

TSUNAMI CHARACTERISTICS

In addition to shaking the land, earthquakes in or near the sea can create extreme waves. Scientists believe that some of the largest waves ever experienced on earth were generated by earthquakes in prehistoric times. In the decade from 1990 to 2000, 14 seismic sea waves hit somewhere in the world, resulting in extensive damage and considerable loss of life. In the Pacific region over the last 2,000 years, nearly 500,000 people have died from tsunami; in contrast, the 2004 Sumatra tsunami in the Indian Ocean alone exerted a death toll now estimated at more than 280,000 persons.[1]

Ironically, the Indian Ocean has had little tsunami activity in recent history, although the circular area south from Myanmar to Indonesia and east and north along the Philippines bounds the Eurasian Plate—a hotbed of seismicity over the last 100 years. There have been nine magnitude 8 earthquakes since 1900. The magnitude 9 Sumatra-Andaman Island earthquake of December 26, 2004, occurred along this fault zone. The lack of tsunami associated with these numerous earthquakes created a false sense of security in the region.

Seismic sea waves—often referred to incorrectly as "tidal waves"—are called *tsunami*, which is Japanese for "large wave in harbor," to distinguish them from tide waves. The same spelling is used for the singular or plural form. Tsunami are quite different from storm-generated waves in that they have long wavelengths between 6 and 300 miles. Recalling from Chapter 1 that "deep" water for a wave is defined as a water depth equal to or greater than half a wavelength, and recognizing that the oceans are generally no deeper than 3 miles with an average depth of around 2.5 miles, it is clear that the entire ocean is "shallow" to most tsunami waves. Hence the speed of a tsunami depends solely on water depth (as in the case of wind waves approaching the shore), but more significantly, the *entire* column of water—from surface to sea bottom—moves in the case of a tsunami. Since the speed of the wave depends only on depth, in the open ocean—where the average depth is 2.5 miles—the speed of the wave is 384 knots (442 miles per hour).

Near an earthquake source in the open sea, the wave height may be only 3 to 6 feet and would not be noticeable to a large vessel. As the wave approaches the shore, things get more interesting. As the depth decreases, the wave slows, causing the wavelength to decrease and the moving mass of water to "pile up," dramatically increasing the wave height. When this massive wave runs up on the shore, it sweeps all before it, carrying boats in the harbor onto dry land and destroying waterfront installations and buildings inland. Some buildings that survive the initial onslaught will be swept out to sea as the water floods back into the ocean. If the initiating earthquake caused the seafloor to rise, the crest of the tsunami will reach the shore first. If the floor sinks, a trough is created, followed by a crest. In this case, the first event on shore is that the water recedes a great distance. Spectators who rush to the seashore or harbor to look at the receding sea find a few minutes later that they are doomed by a great wall of water when the crest arrives (see Figure 13).

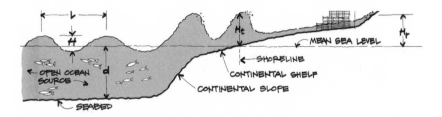

FIGURE 13 Tsunami wave height and run-up.

Geoscientists have recently uncovered evidence that huge sea waves—possibly caused by a massive submarine earthquake, an underwater landslide, or a meteorite or comet impact in the ocean—have swept inland for as much as 3 miles in Australia, inundating hills as high as 200 feet! The evidence for such events comes from places where the impact of giant wave-tossed boulders has left star-shaped fracture patterns in rock faces, where coastal hills have been carved by wave action, or where beach soils and shells have been deposited far inland.[2]

TSUNAMI IN ANCIENT TIMES

Tsunami are known from ancient times—dating back almost 4,000 years in China. In the Mediterranean, perhaps the earliest written record comes from the ancient city of Ras Shamra (Syria). Near the modern city of Latakia, on Syria's Mediterranean coast, was the ancient port of Ugarit. The city was a center of trade for the Minoans and had connections with ancient Egypt. At some time between 1400 and 1200 B.C. the city was destroyed, possibly by an earthquake followed by a tsunami. Archaeologists excavating the ancient ruins have uncovered several libraries containing hundreds of clay tablets written in four ancient languages. They detail trade transactions and legends, and mention the port's destruction by giant waves.[3]

To the west of Ugarit, at nearly the same latitude, lies the island of *Kríti* (Crete), and about 80 miles to the north of Crete is a small group of five islands, collectively called the Santorini Islands, the largest of which is *Thira* (Thera). If you divide the Mediterranean into two halves separated by a line drawn south from the toe of Italy, the island of Crete would lie almost in the center of the eastern half.

Within the Santorini Island group is a gigantic subsurface caldera that seismologists suspect is the remains of a huge volcano that exploded in ancient times—probably around 1600 B.C.—with a force far greater than the one that destroyed Krakatoa. The ancient island of Thera was demolished; the Santorini Islands are the remaining pieces. The Santorini Islands remain active; eruptions have continued—for example, in 1939, 1950, and 1956—ever since.

Along the coast of Crete there is evidence that an ancient tsunami

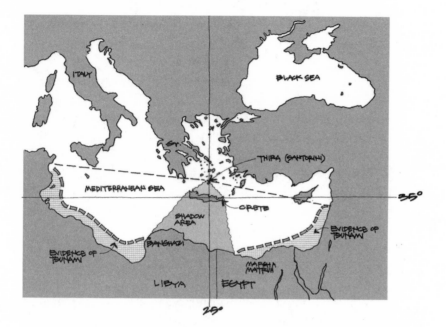

FIGURE 14 Map of Thira Island and vicinity.

struck the island at an elevation of 300 feet. Similar evidence has been found on the coast of North Africa as far west as Tunis and to Ugarit on the east shore of Syria. However, the trace disappears along the North African coast from Banghazi, Libya, to Marsa (Matruh), Egypt—precisely in the "shadow" of Crete (see Figure 14). Such a cataclysmic event was sure to be noted by ancient people all around the Mediterranean, and indeed the ancient Egyptians were aware that something had happened, somewhere. It is possible that this event reappeared in ancient legends as a great flood. Historian Douglas Myles has speculated that the tsunami caused the parting of the waters referenced in the Old Testament when Moses led the Hebrews out of Egypt.

On July 9, 551, in Lebanon, a very strong earthquake—its epicenter offshore from Byblos—devastated Beirut and caused the collapse of many buildings. It was followed by a tsunami in which the sea at first receded around 2 miles and then came roaring back to rush inland. A total of 30,000 persons died in the earthquake and tsunami.[4]

Other early records of tsunami include a number that struck Japan; the disastrous Port Royal, Jamaica, tsunami of 1692; and the Lisbon earthquake and tsunami of 1755.

Japan has a long history of earthquakes and tsunami, dating back at least 1,300 years. In A.D. 869 a tsunami killed 1,000 persons: another in 1293 killed 30,000. In 1611, an 80-foot-high wave entered Yamada Bay, killing thousands. Other notable Japanese tsunami include those of September 20, 1498, in Nankaido, Japan, which killed 26,000; October 28, 1707, in Nankaido, Japan, responsible for 30,000 deaths; and June 15, 1896, in Sanriku, Japan, where a wave 125 feet high claimed 27,000 victims. In the 400-year span from 1596 to 1996, more than 25 major tsunami hit the Japanese islands.[5]

TWO TSUNAMI OF HISTORICAL SIGNIFICANCE[6]

In 1691, William Dampier, the British seaman and explorer cited in Chapter 2, returned to England. Nine months later, he was already anxious to return to the sea, planning to go back to Port Royal, Jamaica, to engage in trade in the Caribbean. Before he could set sail, word reached England that Port Royal had been wiped out by an earthquake and subsequent tsunami. The major buildings had collapsed and some streets had dropped into the sea. Several thousand people had died; the report indicated that the living were too few to bury the dead. No records exist concerning the height of the tsunami that overran the city. However, the wave destroyed most of the shipping in the harbor, and a British frigate was reportedly carried over the tops of two- and four-story houses near the waterfront.

Port Royal represents an important archaeological site because it is in effect a piece of seventeenth-century history frozen in time by the tsunami. In the late 1960s an expedition led by Robert Marx explored the site using accepted archaeological methods. Six months were spent mapping the site, and then the divers began their work. More than 50,000 items were recovered, among them household goods, including silver and pewter plates and cups, glass bottles, jewelry and silver coins, tools, ship's fittings, and many other items. Still, only a fraction of the submerged city was explored and much more remains to be done.

Lisboa (Lisbon), Portugal, was the hub of a great seafaring nation. Early in the morning of November 1, 1755, a huge earthquake occurred somewhere out in the Atlantic to the west of the city. Terrible destruction resulted—so much that although no instruments existed at that time, seismologists have estimated its magnitude as 9.0. Around an hour later, waters in the bay receded and the first wave hit with a height of 50 feet. It rushed inland, completing the destruction caused by the earthquake. It was followed by two more waves. An estimated 60,000 persons died and approximately 80 percent of the city was destroyed.

EXAMPLES OF RECENT TSUNAMI

Following an earthquake on August 13, 1868, a 70-foot-high tsunami swept over Arica, Peru (now part of Chile). The U.S. gunship *Wateree*, a side-wheel steamer, was one of several ships in the harbor that witnessed the town collapse as the earthquake struck. At first the sea receded, causing the *Wateree* to settle on her flat bottom. This perhaps saved her, for when the actual wave came rushing in, she was carried 3 miles up the coast and nearly 2 miles inland—over tall buildings, dams, and trees. As the waters receded, *Wateree* was deposited upright in the desert next to a Peruvian man-of-war, *America*. In this position, with no possibility of returning to the sea, the *Wateree* maintained operations in the desert for a while, assisting in relief efforts while waiting for a U.S. vessel to come retrieve the crew. Since the town was devastated—and much of its population dead—the crew lived on the boat. The crew planted a vegetable garden nearby, and when the captain was "piped ashore," it was on a burro, rather than the captain's gig.[7]

While tsunami are most commonly caused by submarine earthquakes, they can also be caused by landslides, volcanoes, or meteorite impacts in the ocean. Geoscientist Edward Bryant calculates that in the Pacific region during the last 2,000 years, 82 percent of tsunami have been caused by earthquakes, around 5 percent by volcanoes, 5 percent by landslides, and 8 percent by unknown means.[8] The damage caused by tsunami, and the wave heights created, depend not only on the magnitude of the earthquake or other cause but also on the nearshore configuration of the land. Tragic and sometimes strange and miraculous events have occurred, as demonstrated by several major tsunami within the last 60 years.

The April 1, 1946, magnitude 7.3 Alaska earthquake caused a tsunami that struck Unimak Island in Alaska and destroyed a lighthouse at Scotch Cap, killing the crew of five men. The wave also hit a coast guard station on the cliffs above the lighthouse (105 feet above sea level) and heavily damaged it. The lighthouse was a reinforced-concrete structure located on a shelf 46 feet above mean low-water level. The light was 98 feet above sea level. The earthquake occurred at 1:30 A.M. and was strongly felt at both the lighthouse and the station on the cliff above, but a telephone call to the lighthouse crew indicated that they'd had no damage. About 25 minutes later there was a second quake, followed by still more aftershocks throughout the night and next day. What the lighthouse crew did not—could not—know was that the earthquake had generated a tsunami that was now hurtling toward them at about 260 knots. The epicenter of the first quake was later found to be 93 nautical miles distant—the second, even closer. A little after 2:00 A.M. the crew in the upper station heard a roaring noise and then a loud crash described as a "sonic boom." A wall of water rocked the coast guard station and flooded past it. The phones to the lighthouse went dead. Following the initial shock, the crew in the upper station evacuated and moved to higher ground. Looking back, there was no sign of the lighthouse or its powerful light. When daybreak came, they investigated and found the lighthouse destroyed—ripped from its foundation and smashed to pieces. The mangled remains of two crew members were found, the others presumably swept into the sea.[9]

Waves also traveled across the Pacific and about four and a half hours later struck the Hawaiian Islands and other locations, the greatest damage occurring at Hilo, on the island of Hawaii, where 150 people died. The run-up on the Hawaiian Islands ranged from a low of around 10 feet on the sheltered back side of the islands to highs of 52 to 56 feet on the northerly shores facing the oncoming waves.[10] Why was the damage so great at Hilo? To find out, the Army Corps of Engineers built a scale model of Hilo's triangular bay. The model—85 feet wide—

faithfully reflects the bottom characteristics of the bay and its shore-line, including docks, rivers entering the bay, bridges over the rivers, and other important features. Incoming tsunami are simulated by re-leasing water from large tanks. The model tests have demonstrated that due to Hilo Bay's unique configuration, almost any sizable wave that enters the bay will hit downtown Hilo or the coast north of town. Waves are reflected and refracted by the shoreline, and in some cases constructive interference of these waves creates very large waves in the center of the bay.[11]

Following the 1946 tsunami, tsunami warning and prediction cen-ters were established in Hawaii and Alaska. Consequently, when the next Alaskan earthquake tsunami (caused by the March 9, 1957, mag-nitude 8.3 earthquake near the Aleutians, south of Andreanoff Island) hit the Hawaiian Islands, even though the waves were higher, there were no deaths because the warning enabled evacuation of low-lying areas. The tsunami from this earthquake—with a maximum wave height of 12 feet—caused flooding at Laie Point, Oahu, as shown in Plate 9. On Unimak Island, where the Scotch Cap lighthouse was de-stroyed in 1946, waves 39 to 49 feet high were reported.[12]

"The largest earthquake in the world," as it is now being called, occurred on May 22, 1960, near the southern coast of Chile. The mag-nitude of the earthquake, originally reported as 8.3, has been recalcu-lated as 9.5. It generated tsunami waves as high as 82 feet that hit shore 15 minutes later, killing thousands of people along the Peru-Chile coast. The wave also traveled westward 5,400 nautical miles across the Pacific and struck Hilo, on the big island of Hawaii, 15 hours later, killing an additional 61 persons and inflicting extensive damage (see Plate 2). The initial 35-foot-high wave was followed by seven more that arrived every 15 minutes or so. The wave caused damage and 138 deaths in Japan as well. The tsunami impacted the West Coast of the United States but did not cause serious damage or loss of life. Wave amplitudes of 4.6 feet were recorded at Santa Monica. Boats broke loose from moorings, some sinking in Los Angeles and Long Beach harbors. Other vessels were knocked around in San Diego harbor and caused damage. Crescent City recorded a wave height of 12 feet and boat dam-age.[13] (See Figure 15.)

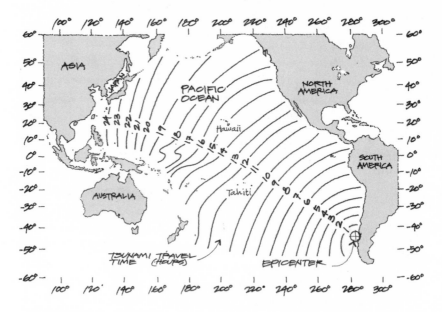

FIGURE 15 *Tsunami travel time in hours, Chile earthquake, May 22, 1960.*

Alaska was struck again on March 27, 1964. The epicenter of the magnitude 9.2 Prince William Sound earthquake was not far from the town of Seward. Anchorage, some 74 miles northwest of the epicenter, Valdez, 56 miles east of the epicenter, and other towns along the coast were hard hit by the intense shaking and by the tsunami that resulted. Seward was struck by a 33-foot-high tsunami, badly damaging docks and harbor installations; the force of the water was so great that boxcars were carried inland. In the source area, the average run-up elevation was 36 feet and the maximum was 220 feet! The earthquake caused the largest true tsunami, 219 feet high, which hit Valdez, Alaska. This wave was 10 feet higher than the tsunami that hit Russia's Kamchatka Peninsula in 1737. A huge area of southern Alaska subsided. Years later, when I visited the area, there were vast areas of dead trees killed by saltwater still standing.[14]

The tsunami also damaged coastal areas in Canada—including Port Alberni—and in Washington and Oregon. Later, the wave struck the town of Crescent City in northern California, killing eight people and inflicting $11 million in damage. It finally reached Hawaii, where the wave height was measured at 14 feet at Waimea and additional damage occurred. There were 131 deaths, 122 of which were due to the tsunami, 5 in Hawaii, 16 in California and Oregon, and the balance in Alaska.

Earthquakes in the Gulf of Alaska have caused tsunami and severe damage, but the magnitude 7.8 earthquake of March 6, 1988, had some strange effects. It was felt in communities all along Alaska's Gulf Coast and produced a small tsunami that measured 1.25 feet at Yakutat, but caused no damage or injuries. The epicenter was in the Gulf of Alaska about 200 nautical miles south of Valdez.

At the same time the 500,000-barrel crude oil tanker *Sansinena II,* under the command of Captain Bent Christiansen, was steaming from Portland, Oregon, to Valdez, Alaska, to pick up a load of crude. Captain Christiansen is now chief port pilot for the Port of Los Angeles. Here is his account of what happened.

"We were on a northwest heading making 16 knots with the wind out of the southeast at 30 knots and a southwest swell of 15 feet. I was on the bridge. Suddenly, without warning, an extremely severe vibration started to shake the entire ship. My first thought was that we'd lost one or several propeller blades. I immediately pulled the throttle back to about 40 rpm, but there was no change in the intensity of the shaking, so I pulled the throttle to FULL STOP. I called the engine room and asked the engineer on watch if he knew what was causing the shaking. He did not know, so next I ran out on the bridge wing to look around. I could see the stack shaking so hard I thought it might collapse. I returned to the bridge and a few moments later the shaking subsided.

"About this time I heard a call over the very high frequency (VHF) emergency Channel 16. It was the *Exxon Boston* calling the *Exxon North Slope* and reporting that she had encountered heavy vibrations, had lost power, and was experiencing some flooding. The *Exxon North Slope* also was without power and called a third Exxon ship, the *Exxon New Orleans,* which turned around and headed back to stand by the *Exxon Boston.* (The *Sansinena* had a steam turbine, no diesel engine in those days, and I believe the same was true of the Exxon tankers.) Meanwhile our radio operator heard about the earthquake from a station at

Ketchikan. We proceeded to check the deck and the engine room and found no signs of damage, so I gradually resumed speed. I called the *Exxon North Slope*, gave our position, and offered to assist if needed. They responded that they were restoring power and did not need any help at this time. The *Exxon Boston* reported that the flooding was under control and the *Exxon New Orleans* was now standing by. While this was happening we felt the first of several aftershocks, leaving no doubt that it was an earthquake we'd felt. We resumed course to Valdez, where eventually all vessels arrived without further incident.

"I plotted the position of the earthquake and the positions of the three Exxon tankers and found that they were about 35 nautical miles southwest of our position. That put them about halfway between us and the earthquake epicenter. I would never have believed that we could feel such a sharp, jarring motion transmitted through that great depth and distance of water."

I asked Christiansen if he had noticed any change in the sea state following the earthquake. "No," he said. "If there was a wave, it was too small for us to notice. But that hasn't always been the case. In the five years I was master on the *Sansinena II*, we ran into big waves several times. These were 75 feet high and covered the entire 800-plus-foot-long deck with green water. Waves hit the bridge as well. At such times about all you can do is heave to. You can't help but have a number of concerns, foremost being not to lose power. I don't know if those are what you would call extreme waves, but for me they were pretty extreme."

Coincidentally, as I thanked Captain Christiansen for talking with me, he mentioned that he was on his way to pilot the *Lane Victory* back into Los Angeles Harbor. I told him to be sure to say hello to Ernie Barker (see Chapter 8), since steam ships and big waves are something that he and Ernie have in common.

The list could go on and on, but I will mention just a few others to indicate the geographical distribution of tsunami occurring in a short time period. On September 1, 1992, a magnitude 7.0 earthquake occurred off the coast of Nicaragua, causing a wave with a run-up of 33 feet. Three months later, on December 12, 1992, very nearly on the opposite side of the globe, a magnitude 7.5 earthquake struck near

Flores Island, in the Sunda-Banda Island group (Indonesia), where the average run-up was 16 feet, with a maximum height of 65 feet, causing 2,080 deaths. On July 12, 1993, a magnitude 7.8 earthquake in the Sea of Japan and the resulting tsunami hit Okushiri Island with an average run-up of 33 to 49 feet and a maximum of 98 feet. It killed 185 people and caused extensive property damage.

TSUNAMI CAUSED BY LANDSLIDES AND VOLCANOES

Earthquakes can cause large waves by means other than displacement beneath the ocean's surface. On July 10, 1958, a magnitude 7.9 earthquake occurred in southeast Alaska, not far from the current site of Glacier Bay National Park, causing landslides, submarine slides, and icefalls from glaciers and producing six separate wave events. Since the earthquake occurred in a remote area, there was little damage or loss of life, but it did result in several amazing survival stories.[15]

The earthquake dislodged a mammoth wall of rock and pieces of ice from the glacier in the headlands of nearby Lituya Bay, creating a huge wave that roared through the bay and out into the open sea of the Gulf of Alaska at a speed of 80 to 110 knots. Some would call this a "splash," rather than a tsunami. Whatever you choose to call it, to the crews of the three boats anchored in the bay that day, it was the biggest wave they'd ever seen.

The salmon troller *Edrie*, with two crew (Howard Ulrich and his six-year-old son, Howard Jr.) on board, was anchored inside the bay; two other boats were anchored near the entrance, at a place called Anchorage Cove. Ulrich heard a roaring noise and saw the wave coming as it broke around Cenotaph Island, a 320-foot-high island in the middle of the bay. The wave was steep and 66 to 98 feet high as it approached his boat. He frantically tried to maneuver, but the wave picked up *Edrie*, swept it up and over dry land, and then by a random chance of fate dropped it back into the bay. The other two boats (*Badger*, with Bill and Vivian Swanson aboard, and *Sunmore*, with Orville and Micki Wagner aboard) were swept out to sea over the tops of trees on a spit of land at the entrance to the bay; both sank, but the Swansons scrambled into a dinghy and survived. The Swansons reported that the wave first hit the southern side of the bay near Mudslide Creek, where the water height reached 656 feet; then as the wave passed Cenotaph Island, it cleared trees to a height of about 160 feet, before hitting them and tossing their boat over the sand spit at the mouth of the harbor.

Aerial photographs taken after the earthquake show the mountains on the north and south sides of the bay swept clean of trees and vegetation. Water surged to a height of 1,700 feet as it poured over the fingers enclosing the bay, stripping off the trees and topsoil down to bare bedrock. Geological records and examination of tree-growth rings show that this was not the first time a giant wave has coursed through Lituya Bay; apparently large waves occurred in 1854, 1874, 1899, and 1936. Recently, scientists have developed various wave models for the bay to predict what wave height could occur under various scenarios. They conclude that a 1,700-foot-high wave is indeed possible.[16]

Volcanoes are another source of tsunami; at least 92 cases have been documented.[17] Of these, the most famous is the eruption of *Krakatau* (Krakatoa) on August 27, 1883. Following a series of lesser eruptions and explosions for several days preceding the 27th, the volcano finally destroyed itself in an immense explosion that literally echoed around the globe. In the Sunda Straits nearby, ships were lost, 165 villages were destroyed, and more than 36,000 people were killed along the coasts of Java and Sumatra, mostly by two extreme waves that followed when millions of tons of debris were dislodged into the sea. These tsunami had periods of one to two hours. Many towns were only about 33 miles from Krakatoa, so there was no way a warning could have been issued. In addition, since the volcano had been erupting periodically for days, people became complacent. But when it finally blew up, within minutes the town of Merak on the island of Java was hit by a wave 100 feet high. In Merak the wave destroyed stone buildings on top of a hill that stood 115 feet high. Lighthouses toppled, a naval vessel was picked up in the harbor and carried several kilometers inland, and from the high water marks, it appears that the largest wave ran up at least 133 feet when it hit the shore.[18]

TSUNAMI WARNING SYSTEMS

In its most elementary form, a tsunami warning center has three major components: seismographs to detect if an earthquake capable of causing a tsunami has occurred, tide stations that monitor water level and indicate whether or not a tsunami wave has been created, and a means of disseminating the warning via a variety of different, robust communications routes to civil defense or disaster management authorities. A number of nations operate regional tsunami warning systems. However, of the three major oceans, only the Pacific Ocean has an integrated multinational tsunami warning system. Twenty-six nations in or bordering on the Pacific Ocean participate in an International Coordination Group formed under the auspices of the United Nations Educational, Scientific, and Cultural Organization's (UNESCO's) Intergovernmental Oceanographic Commission.

The nerve center for the Pacific Ocean warning system is a small concrete block building located at Ewa Beach, a short distance west of Pearl Harbor on the south side of the island of Oahu. Following the 1946 Alaskan earthquake, the U.S. government established the first elements of a tsunami warning system that linked the mainland and the Hawaiian Islands. Then, after the Chile earthquake in 1960, the system was expanded to cover all of the countries along the edges of the Pacific Ocean, the so-called Ring of Fire. I visited the center to get a first-hand look at its operation. I was met by tsunami experts Barry Hirshorn and Stuart Weinstein, who kindly gave me a detailed briefing on how the system operates. Hirshorn and Weinstein live in houses next door to the center. They each carry not one, but *two* pagers, set to go off automatically if an earthquake capable of causing a tsunami occurs anywhere in the Pacific. Even if they are not already in the center, within 90 seconds they will be assembling the data to decide if a tsunami bulletin should be issued.

There are four levels of bulletins issued by the center upon detection of an earthquake of magnitude 6.5 or greater:

• A tsunami *information* bulletin: This is a message to advise all participants in the warning system that a major earthquake has occurred in the Pacific. The bulletin will also advise that it does not appear that a tsunami has occurred or that the possibility of a tsunami is being investigated. The reason for this uncertainty is that depending on the location of the earthquake, more time and data may be needed to make the determination.

• A fixed tsunami *warning* for all coasts within 1,000 kilometers (620 miles) of the epicenter of a potentially tsunamigenic earthquake.

• A regional tsunami *warning or watch* bulletin: If a tsunami appears possible, all areas within a three-hour tsunami travel time will be placed on a warning status and areas within a three- to six-hour travel time will be alerted to a "watch" status. The purpose of the "warning-watch" system is to provide as much advance warning as possible to enable local areas to mobilize warning and evacuation systems, but at the same time to try to minimize the number of false alarms.

• A Pacific-wide tsunami *warning* bulletin: Once the presence of a tsunami with destructive potential that goes beyond a local region has been confirmed, a broad general warning is issued with hourly updates to all areas with coastal populations.

While the Pacific Tsunami Warning Center issues tsunami bulletins for the entire Pacific region, there are a number of important regional systems that cover local areas and share data with the center. An example is the regional system of Japan. Due to Japan's long history of devastating tsunami, this country has the most advanced system in the world, consisting of more than 1,500 seismometers and more than 500 water-level gauges. The Japanese system includes a number of underwater seismometers. The location of an earthquake can be pinpointed in Japanese waters and tsunami warnings can be issued within a minute. Timing is especially important for Japan because of the large number of destructive earthquakes in nearby waters that have caused tsunami. The travel time for these tsunami is so short that warnings must be nearly instantaneous to be effective.

Other regional systems include those in French Polynesia (Tahiti), Russia, Chile, and Australia. For Alaska and the West Coast of the United States there is a regional warning center located in Palmer, Alaska, about 45 miles from Anchorage. Besides serving as the warning center for the entire Pacific Ocean, the Pacific Tsunami Warning Center performs a dual role as the regional center for the Hawaiian Islands and all other U.S. interests in the Pacific.

A decade or so ago it took about 60 minutes for the center to evaluate the data, determine where the earthquake was located, and issue a warning for an event occurring anywhere in the Pacific Basin. Regional warnings for earthquakes occurring in the Hawaiian Islands could be issued in about 15 minutes at that time. Today, with more instruments sending data to the center, faster computers, improved computer software for earthquake analysis, and models for tsunami wave propagation, these times have been reduced to 20 to 25 percent of the previous values. A warning for a major tsunami from an earthquake occurring in the Pacific Basin now takes about 15 minutes; a regional warning can be issued in 3 minutes.

The heart of the detection system is the Incorporated Research Institutions for Seismology (IRIS) global seismic network, an array of approximately 140 broadband seismometers located throughout the world. In addition to those installed and maintained by IRIS and additional instruments on the U.S. mainland installed by the United States Geological Survey (the USGS National Seismic Network), the center automatically receives data from seismometers operated by some of the other member countries. The heart of the detection system for locally generated tsunami in Hawaii is the center's local network of seismometers and water-level instruments and the USGS Hawaiian Volcano Observatory seismic network on the island of Hawaii.

When an earthquake occurs, the first step is to get a rapid estimate of its location by examining the readings from several different seismometers; then, by knowing how fast seismic waves travel through the earth, the distance to the earthquake can be calculated. By using several such readings and a process of triangulation, the location of the earthquake is established. The more seismometers, the more quickly an accurate determination can be made. Unfortunately, the Pacific Ocean encompasses a vast area, and the center would like to see more seismometers located within it. To provide better coverage, a number of these should be underwater instruments like those the Japanese have, but they are very expensive.

The next step is to determine the earthquake's magnitude. This is not a simple matter of reading a gauge; it requires a computational process. The initial estimate is usually refined later as more data are collected. Historically, tsunami are rarely caused by earthquakes with magnitudes less than 7.5. But to stay on the safe side, any earthquake with an apparent magnitude of 5.5 to 6.0 will trigger the evaluation process.

The second element in the detection system is composed of water-level measuring devices. There are two basic types: shore-based systems and deep-water systems. Fortunately, there are a large number of automatic tide stations around the Pacific Rim. These stations transmit water-level information to the center via satellite. The warning system employs eight deep-water buoys to monitor deep ocean waters. These buoys consist of pressure sensors placed on the bottom of the ocean. Their sensitivity is such that they can detect a tsunami wave from as little as 0.4 inches in height to as much as 40 inches. (Remember, tsunami waves are not very high in deep water.) The bottom sensor (which may be in waters as deep as 3 miles) transmits data acoustically to a tethered buoy on the surface. The buoy in turn transmits the information to a satellite, which relays it to the center. The deep-water instruments constitute an important part of the detection system and several dozen more are needed. However, they are expensive to install and maintain and cost around $250,000 each. In addition to these instruments, the center receives sea level data from more than 100 stations operated by member countries.

At the shoreline, the problems are different than in the deep ocean where the buoys are. First, it is difficult to get accurate readings, since local effects (run-up or seiches) can make a given wave 3 feet high in one location and 30 feet high in another. In a major tsunami, the shoreline instruments in the vicinity will be driven off-scale or destroyed, rendering their readings useless, though they may still be useful in helping coastal observers provide a warning to adjacent areas.

Data from all of the seismographs are monitored continuously on the center's computers. Data from ocean- and shore-based tide instruments are transmitted by satellite to the center. Sea level measurements are made every two seconds, and then averaged over three or four minutes and transmitted to the center every three or four hours. The center's computers actuate pagers as soon as an earthquake occurs or if large water level amplitudes are observed.

Once it has been determined from the seismometer data that a tsunami may have been formed, the center issues a bulletin and then scans the water level gauges in the vicinity of the earthquake to determine if a wave has been detected and, if so, how large it is. The bulletin is issued before a tsunami has been observed, by estimating potential tsunami travel times based on bathymetry and a tsunami propagation model developed by Paul Wessel at the University of Hawaii at Manoa. Countries that lie in the potential path of the tsunami now have adequate warning time to evacuate low-lying coastal areas, unless they are located very close to the epicenter. While this procedure does not eliminate property damage, it has drastically reduced the number of deaths attributed to this type of extreme wave. With this information—a combination of nearshore tidal data, deep-ocean data, and the size and type of the earthquake—accurate forecasts of a tsunami's danger can be made. Here again, the speed and accuracy of this determination are influenced by how many instruments detect the wave and their location relative to the earthquake. Due to a lack of coverage, there are areas of the Pacific where warnings may be delayed or where the center is unable to determine that the danger is past.

TSUNAMI INTENSITY SCALE

Tsunami intensity and damage potential are measured by the Modified Sieberg Tsunami Intensity Scale.[19]

1. *Very light.* Wave so weak as to be perceptible only on tide gauge records.

2. *Light.* Wave noticed by those living along the shore and familiar with the sea. On very flat shores generally noticed.

3. *Rather strong.* Generally noticed. Flooding of gently sloping coasts. Light sailing vessels are carried away on shore. Slight damage to light structures situated near the coasts. In estuaries, reversal of the river flow some distance upstream.

4. *Strong.* Flooding of the shore to some depth. Light scouring on man-made ground. Embankments and dikes damaged. Solid structures on coasts injured. Large sailing vessels and small ships drift inland or are carried out to sea. Coasts are littered with floating debris.

5. *Very strong.* General flooding of the shore to some depth. Quay walls and solid structures near the sea are damaged. Light structures

are destroyed. Severe scouring of cultivated land; coast littered with
floating items and dead sea animals. With the exception of large ships,
all other vessels are carried inland or out to sea. Big bores in estuary
rivers. Harbor works damaged. People drown. Wave accompanied by
strong roar.

6. *Disastrous.* Partial or complete destruction of man-made structures for some distance from the shore. Flooding of coasts to great
depths. Large ships severely damaged. Trees uprooted or broken. Many
casualties.

For major earthquakes that occur within a few hundred kilometers (100 miles or so) from a coast, if a tsunami is created, the travel
time of the wave is such that it will hit the coast at any time from
within a few minutes very near the event to as much as 40 minutes
farther away from it—the Sumatra earthquake on December 26, 2004,
is an example. In such places there is a high risk of loss of life unless
immediate steps to evacuate are taken. Even if the center issues a warning, there is very little time to evacuate in this case. Once the tsunami
warning is sounded, residents need to know they must immediately
walk inland to higher ground. Coastal communities need to understand that if they feel the ground shake, there is probably a large offshore earthquake occurring and they should *immediately* evacuate and
not wait for a tsunami warning.

Barry Hirshorn is enthusiastic, articulate, and very committed to
his work. We were in the Pacific Tsunami Warning Center control
room, surrounded by racks of computers, monitors, plotters, electronic
instrument racks, and large-scale maps, looking at displays that showed
tidal measurements from around the Pacific, when he brought up the
subject of evacuation. He told me: "If you feel the ground shake, or see
the ocean recede or behave in an unusual way, get out of there. Don't
drive—leave the Mercedes or that brand new SUV behind! Immediately walk inland. A fast walk of only 10 to 15 minutes will probably be
enough to save your life. Remember, there will be multiple waves, and
it may not be safe to return for hours."

The Lituya Bay wave described earlier is another instance in which
a warning might not be possible. If the earthquake is too small to trigger the Pacific Tsunami Warning Center's pagers, but has caused a large

submarine landslide, the center might not detect this as a tsunami source. The scientific community is working on this problem and hopes to use global positioning satellites to see these submarine landslides even if there is no earthquake to trigger them. To address this problem in the interim, the center has written software that triggers the pagers if large amplitudes are detected on certain coastal water level gauges.

Finally, as Hirshorn pointed out to me, the Pacific Tsunami Warning Center is analogous to one leg of a three-legged structure. The other two legs are emergency management and public education and awareness. Without the other two "legs," the center's warnings are of little value. Once the center has sent out a warning, communities likely to be affected by the tsunami need to implement predetermined emergency procedures. Typically these will involve mobilization of police and other emergency services, beach clearance, sounding of sirens or alarms, and traffic control to permit prompt evacuation of low-lying areas.

Public education and awareness are essential. The public needs to be informed in order to evacuate promptly in the event of a major offshore earthquake or if the ocean is observed to suddenly recede. If a warning is given, people need to know the shortest route to high ground—and, after a wave has hit, to stay clear of low-lying areas because there are very likely to be many more waves on the way. There can be waves for hours after the initial wave hits, and often the first wave is not the largest.

The importance of public education was proven dramatically by the Papua New Guinea tsunami of July 17, 1998, which slaughtered 2,500 people—75 percent of the population in the coastal areas hit by the tsunami. When the next tsunami occurred, the death toll was drastically reduced—because a public education campaign had informed people of the warning signs and the steps to be taken to evacuate. Also, during the Southeast Asia tsunami of December 2004, fatalities were low on several Indian Ocean islands simply because the village elders saw the ocean waters withdrawing from the shore; recognized, thanks to village lore, that this phenomenon heralded a tsunami; and warned the populace to withdraw to high ground. Others were not so fortunate.

7

The Southeast Asia Tsunami of December 26, 2004

There was no tsunami warning system in the Indian Ocean in 2004, although experts had recommended that one be installed. When the December 26, 2004, earthquake was detected, some time elapsed before its tsunami potential could be assessed. Once the risk was realized, the Pacific Tsunami Warning Center in Honolulu tried to communicate with Indonesia and other countries thought to be in the path of the tsunami. But there was no preestablished system for notification, no provisions for issuing a warning even if notification had been made. The effort was too little, too late, and thousands of lives were needlessly lost. At the northwest end of Sumatra, 158 miles from the epicenter of the earthquake, the window for notification was perilously short. It took only about 20 minutes for the first tsunami wave to hit land. At that time, on the afternoon of Christmas Day, Hawaii time, the Pacific Tsunami Warning Center was still analyzing the data in an effort to determine just how big the earthquake was. Unfortunately, it was much larger than originally thought. By then, it was too late. The damage reports were already on television around the world.

The television images are unforgettable. The waves—not one, but several—roll in from the ocean, across beaches and manicured lawns, sweeping all before them. A jumble of beach umbrellas, rattan lawn chairs, towels, and potted plants are borne by the onrushing water. The water courses through open-air restaurants, flows across hotel lobbies and verandas, and pours in dirty waterfalls down staircases. As riveting and terrible as these images are, they are not the worst. Next we see people running frantically, heads turning, looking back over their shoulders at the churning muddy water that is gaining on them. Then, looking closer, we see someone struggling in the maelstrom, buffeted by the tumultuous seas; we see another man clinging desperately to a palm tree, only to lose his grasp and be sucked from sight by the roiling waters; another, carried toward a building, is eventually rescued. (See Plate 10.) Later, we see more, so many more that our senses are confounded; it does not seem possible, such death and destruction—the swollen corpses; mothers crying over infants; the young, so lifelike in death.

It was the worst natural disaster most of us have ever seen or could imagine seeing. It was the Southeast Asia tsunami of December 26, 2004, a definite 6 on the tsunami intensity scale. (Note: Worst is a relative term. Eight months later, almost to the day, the world recoiled to the vision of another terrible disaster: Hurricane Katrina striking New Orleans and the Gulf cities of Louisiana and Mississippi. The cause was different, but the devastation had a terrible familiarity.)

The tsunami was triggered by a massive magnitude 9.0 Sumatra-Andaman Islands earthquake in the Indian Ocean near the northwest end of Sumatra. Scientists believe that more than 600 miles of ocean bottom ruptured and heaved upward as much as 66 feet in a process seismologists call *subduction* when the Indian Plate, moving northeast, slid under and lifted up the Burma Plate. Gradual movement had been proceeding for years in this area, at the rate of about 2.5 inches per year, causing stress to build at the junction of the two tectonic plates, until finally the submarine rock fractured. When this occurred, the locations of nearby islands and the tip of Sumatra were shifted by the earthquake. An earthquake of this size was bound to cause severe structural damage to buildings within a radius of 120 miles, and within seconds, buildings in Sumatra and adjacent areas started collapsing.

It is doubtful that any warning could have saved lives in Sumatra and at other points close to the epicenter, since the tsunami took less than an hour to reach them. In deep water, the tsunami traveled at around 384 knots, but slowed to around 174 knots in the shallow waters near the shore. West Sumatra was hit by three waves, the second reportedly the largest. Extreme wave heights reached 90 feet in some locations, but only 40 feet on the north coast. On the other side of Sumatra, seven waves were reported, resulting perhaps from the primary waves striking Malaysia and reflecting back. Traveling east and northeast in the Andaman Sea, the waves struck Java and Malaysia, rolled ashore at beach resorts in southern Thailand, and hit Burma and Bangladesh. The low-lying Andaman and Nicobar Islands close to the epicenter were hit hard.

Waves propagating west and northwest crossed the Bay of Bengal in around two hours and hit the east coast of India and ravaged Sri Lanka. Remarkably, a passing satellite happened to catch the image of the tsunami as it crossed the Bay of Bengal on its way to Sri Lanka. It was a fast-moving wave about 28 inches high, with a wave front of ever-broadening circumference. When it reached Sri Lanka, it piled up to a height of 20 to 25 feet. It was not just the height that counted in this tsunami; the earthquake was so large, and the surface rupture so great, that a huge volume of water was set in motion. Consequently, when the waves hit shore, the massive flood of water behind the wave kept coming and coming, rushing far inland. There are many graphic reminders of the volume of water: debris left on second-floor balconies far inland, bodies deposited in the tops of tall trees, and cars and boats left high and dry. Then, as the waters receded, the carnage was doubled or perhaps tripled, as the waves swept walls of debris back toward the ocean. Buildings that might have survived the initial onslaught in a weakened state were torn apart when the waters returned from the opposite direction.

In Sri Lanka, a passenger train running along the coast stopped because of the flooding. A number of passengers stayed on the train; other people jumped on the train to escape the water. The next wave rolled the train, and hundreds of persons were trapped and died.

Exiting the Bay of Bengal, the tsunami next rolled westward across the Indian Ocean, striking the Maldives three hours later, then Madagascar, and the east coast of Africa—Somalia, Kenya, and Tanzania. Since it took seven to eight hours to cross the Indian Ocean, Africa had some warning. But even here, 3,000 nautical miles from the epicenter, people died.

Out of this devastating natural catastrophe have come incredible stories of heroism and survival. Some of the most unique witnesses to the tsunami's fury were in the ocean—scuba divers underwater off the coasts of Thailand or Sumatra, who felt the turbulence of the passing wave, and in some cases rose to the surface to find their dive boats gone. Or fishermen at sea, who, if they were in deep enough water, survived the passing wave, only to find when they returned to port that they could not recognize anything, so complete was the destruction.

DIVING UNDER A KILLER TSUNAMI

A minute before 8:00 A.M. on December 26, 2004, guests in the tourist hotels on Patong Beach, Phuket Island, Thailand, were wakened by a tremor from a distant earthquake, followed by a second one about 20 minutes later. No one thought much about it. It was also felt on Koh Phi Phi, a small island about 24 nautical miles east of Phuket. Gene Kim and Faye Wachs, in the process of boarding their dive boat in Ton Sai Bay, Koh Phi Phi Island, did not notice the slight ground movement. They were approaching the end of a two-week vacation in Thailand, a vacation that had taken them from the mountains in the north to the idyllic beaches at Thailand's southern tip. They came to Phuket on December 24, spent the night in a hotel on Patong Beach, and then caught the first ferry to Koh Phi Phi on the morning of December 25. Upon arrival they rented scuba gear and made an exploratory dive, telling the dive master that if it was a good trip, they'd come back the next day for a longer "three-tank" dive. It was a good trip—close-up looks at leopard sharks and other fish, and Kim and Wachs were ready for more. The dive master promised even more spectacular experiences for the three dive sites to be visited on the next day. Little did they know that the "spectacular experience" would be their salvation.

On board the dive boat on December 26 with Kim and Wachs were two Thai boatmen, an Israeli dive master known to them only as

Yavit, a Swedish tourist known only as Olsen, and a second dive master known only as Erik. Kim and Wachs had with them the bare essentials for their dive as well as a camera and a wallet containing some cash and credit cards. Their passports, airplane tickets, and other belongings were back in their bungalow at the Koh Phi Phi Cabana Hotel. The dive boat headed northwest to the first dive spot, a wrecked interisland catamaran ferry 280 feet long known as the *King Cruiser*. It had run aground on Anemone Reef in 1997 and now sat on the reef in water 98 to 115 feet deep.

As Kim, Wachs, and the others were boarding their dive boat, approximately 310 nautical miles southwest as the dolphin swims, deep in the ocean along the Sunda trench, the seafloor ruptured—the Burma Plate on the east side rising as the Indian Plate on the west side pushed below it. The fault break propagated rapidly northwest toward the Nicobar Islands, moving at a variable speed later estimated to average 62 miles per minute. Within three minutes it traveled the 180 miles to reach Great Channel, the body of water that separates Banda Aceh, at the northern tip of Sumatra, from Great Nicobar Island.

As the dive boat prepared to leave the dock at around 8:15 A.M., the first tsunami waves hit Indonesia. Waves as high as 49 feet virtually obliterated Banda Aceh, at the northern tip of Sumatra. As the fault movement propagated north, new waves were produced. No longer blocked by the Sumatra mainland, these waves had nothing but 300 nautical miles of the open waters of the Andaman Sea between them and Thailand's west coast and jewel-like offshore islands. The new waves, generated as the fault break propagated north, were accompanied by earlier waves that diffracted around the northern tip of Sumatra to hit Thailand and streamed down the Strait of Malacca to hit Malaysia. In the shallower waters of the Andaman Sea, the waves slowed to perhaps 174 knots.

Meanwhile, the dive boat headed northwest, unaware of the cataclysmic sea change bearing down on it. Around 10:00 A.M. the dive boat reached the buoy that marked the site of the wreck and the divers entered the water, dropping beneath the surface along the line attaching the buoy to the wreck. The dive boat—the two crew relaxing on board—drifted on the still waters above the wreck.

A few minutes earlier, those on the beach at Phuket were amazed to see the ocean withdraw a long distance from shore, stranding some boats and leaving flopping fish behind. Some curious spectators walked out to investigate; others turned and ran for high ground, screaming warnings to anyone who would listen. Three waves overran the resort areas including Patong Beach. After steamrolling over Phuket Island, the waves continued northeast to Koh Phi Phi Island. The island is shaped like a lowercase backwards letter "h," with the long part of the h on the right-hand (east) side and the short part on the west. The long and short parts of the h are in reality two islands connected by a narrow, low peninsula. (See Figure 16.) The waves came in from the west, hit the beaches on the top center of the h, roared through Phi Phi village, and crashed into Ton Sai Bay on the other side of the peninsula. Most of the village buildings, including Kim and Wachs' bungalow,

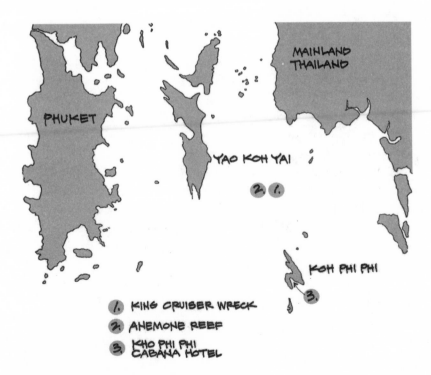

FIGURE 16 Map of Koh Phi Phi Island and vicinity.

were ripped apart and carried across the peninsula and into Ton Sai Bay. Anyone caught in the buildings as the two waves carried the rubble first in one direction and then back again either died or suffered frightful injuries.

On the dive boat, Kim and Wachs were unaware of the impending catastrophe. What happened next is best described in their own words, as they told me the story at the California Yacht Club in Marina Del Rey, California.

Kim: "We descended in two groups—Faye, me, and dive master Yavit as one—and Olsen and Erik as the second group. Yavit had advised us that if strong currents were encountered not to try to swim against them. He gave us hand signals for changing direction, surfacing, and so on. We descended to around 60 feet and proceeded to swim along the side of the wreck. I don't recall seeing any fish, but didn't think about it at the time. We swam to the other end of the vessel and there encountered a strong current. The dive master gave us the signal to turn and go back the way we'd come. As we were returning, the water suddenly became quite turbulent. The dive master was next to Faye, and he grabbed her shoulder and gave the hand signal for an emergency ascent. At this point it was around 10:20 and we were about 10 minutes into the dive. They were a short distance from me, but I saw the signal. The water suddenly became very cloudy—I'd describe it as a complete whiteout. As visibility went to inches, I found myself being tumbled violently, as if in a giant washing machine."

Wachs: "The visibility was so bad that I could barely see the dive master, even though he still had a hand on me. I felt myself being tumbled around and at first I did not know which way was up. I inflated my buoyancy vest, but at first nothing happened. I looked at my dive computer and saw that it indicated a depth of 135 feet. I don't know if this was the real depth or reflected a wave passing above, pressing me down. Finally the vest buoyancy took effect and I started to surface. I went up with the dive master; we stopped at 60 feet for a few minutes, then at 30 feet for a few more minutes, and then surfaced. There was no sign of Gene. I checked my dive computer and saw that 28 minutes had elapsed."

Kim: "I felt myself being sucked downward and tumbled around.

I was initially pulled away from the side of the wreck and then slammed against it, hitting my head solidly on the side of the sunken vessel. I was spun around down to about 120 feet, trying to see which way my air bubbles were going so I could figure out how to get to the surface. Finally I worked my way up into clearer water and miraculously found myself right next to the buoy mooring line we'd used in our descent. I went up to 30 feet and hung there a few minutes trying to figure out what was going on. When I looked up I could see two figures hanging on the line. The line was stretched out at a sharp angle and divers' bodies were stretched out horizontal, indicating the presence of a very strong surface current. I surfaced, hoping to find Faye and Yavik, but discovered it was Olsen and Erik. The dive boat was not in sight. I considered going back down to look for Faye, but then saw that the visibility would be essentially zero once I got down to 30 feet and realized it would be dangerous and probably futile. Meanwhile, one boat appeared in the distance, going away from us, and then a second boat appeared several hundred yards away coming toward us—our boat. We were assisted back in the boat by the crew. The crew spoke no English, so questions were not possible. Reading their body language, the crew did not seem to indicate anything out of the ordinary. I can only assume that the boat was pushed away from the buoy by what appeared to them to be a large swell. A short time later, two heads were seen bobbing some distance away and to my great relief, we got Faye and Yavit into the boat."

Wachs: "I could tell that Yavit was shaken by the experience—said he'd never seen anything like it before. No one had an explanation other than it seemed like a freak current. It was a full moon, high tide, and so we thought it might have had something to do with that. At the moment we had no premonition or thought that it might have been a tsunami."

Kim: "The dive master suggested that we move to the second dive spot, a place on Anemone Reef, where it was more sheltered and there were no currents. We went there, and Yavik went in the water to check it out. He emerged immediately and said the current was too strong, so we aborted the second dive."

Wachs: "We went to the third spot, called Shark Point, supposedly more sheltered. Here again we encountered unusually strong currents. I noticed that the fish behavior seemed strange. Small puffer fish—fish that we usually encountered solitary—were huddled together in groups. The fish seemed unusually skittish and seemed to be schooling tighter. We also cut this dive short. When we were back on the boat, we decided we'd had enough, and elected to return to the harbor early. It was now about 1:30 P.M. The dive master offered us a dive the next day at a location on the other side of the island where we would be sheltered against any turbulence, since this trip was being cut short. We agreed to his plan."

Kim: "It took us about an hour to get back to Ton Sai Bay. As we got closer, we began to see a lot of floating debris that wasn't there when we'd left the harbor in the morning. Our first reaction was anger that someone had dumped a load of trash into the bay. Then the objects began to take on the appearance of stuff that wasn't trash—a floating chair, a TV, a six-pack of water in plastic bottles—unopened—other things. Then we thought that maybe there'd been an explosion or a shipwreck. A jet aircraft (fighter plane) came over, flying very low. This was quite unusual and was the first indication that something might be terribly wrong. Then Olsen got a text message on his cell phone from his wife. It said 'Big catastrophe!' He called back, and she told him there'd been a tsunami. Suddenly it dawned on us what had happened, but we still had no idea about the earthquake or how extensive the damage was."

Wachs: "Until we started seeing the floating bodies."

For the next several hours the dive boat remained in the bay, unable to dock due to the floating debris and damage to the pier. Meanwhile Kim and Wachs assisted numerous fishing boats in the bay by pointing out locations where bodies were floating so the smaller boats could recover them.

Finally ashore at around 5:30 P.M., Kim and Wachs went to their hotel and tried to find their bungalow. Destruction was so complete it was difficult to identify their bungalow among the ruins. Eventually they located what they thought was theirs and confirmed it by the fact that nearby there was a door with their room number. All of their belongings were gone except the few things they'd taken on the boat with them. Passports, tickets, and everything else were lost, although Faye later spotted a pair of her shorts in the mud some distance from the wreckage. Still, they realized their good fortune when they saw their bungalow and the injured people. If they had stayed in the hotel, it is doubtful they would have survived. (See Plate 11.)

Kim and Wachs immediately began assisting the victims—Kim helping excavate a man trapped under building debris, Wachs assisting with the injured. Two vacationing doctors set up an improvised treatment area. There were a lot of badly injured people, severe lacerations, compound fractures, saltwater ingestion. Later, Thai helicopters starting flying in. The injured were triaged, and those needing treatment were loaded on improvised stretchers (usually doors) and carried to the helicopters for evacuation to a hospital. This went on until late in the evening. Finally, at 1:00 A.M., Kim and Wachs were exhausted. They went to one of the two hotels still standing, found an empty room on the third floor, and slept for a few hours.

In the morning they were able to look around and see the extent of the devastation for the first time. They could see the path where the waves had come in over the beaches on the west side of the island and literally swept the village across the narrow peninsula, grinding up the buildings and dumping the rubble into the bay. Kim said that the waves were 30 to 35 feet high on the basis of the waterline in one of the two hotels (reinforced-concrete construction) that survived. He also saw a small powerboat impaled upside down in the top of a palm tree about 35 feet high. Virtually all of the single-story buildings in the area were demolished; just the two reinforced-concrete hotels remained standing.

Kim and Wachs continued to assist the injured. Ferries started arriving to evacuate tourists to the mainland at Krabi. By now they were both suffering from cuts and abrasions on their feet and legs from carrying stretchers through the rubble, and decided it was time to leave. At that point the Thai military had nurses and doctors in place and was managing the evacuation of injured. Kim and Wachs spoke highly of the response of the Thai government in helping the tsunami victims and tourists alike. Still in their swimsuits, they were taken by ferry to Krabi, then flown to Bangkok and finally back to the United States.

They told me that they plan to return to Koh Phi Phi at some point, both to pay homage to those that were lost and to reunite with some of the Thai people with whom they worked and who befriended them during the ordeal.

During my visit to the Pacific Tsunami Warning Center, I asked Hirshorn what new information had come forth concerning the December 26, 2004, Sumatra-Andaman Islands earthquake. The earliest analyses indicated an earthquake of magnitude 8.0, later revised by the center to 8.5. Hirshorn said that it has now been established that the earthquake magnitude was in the range of 9.0 to 9.3, making it one of the two or three largest earthquakes in recorded history. The length of the rupture zone was in excess of 740 miles, with a width around 30 miles. On the basis of the early data, it appears that the southernmost 250 miles of the fault produced the greatest movement, and it was this section that generated the tsunami. The total earthquake duration was on the order of 10 minutes.

The tsunami is unique because the wave was detected in midocean by a satellite. When the *Jason 1* satellite passed over the Indian Ocean on December 26, 2004, it recorded a rise in the ocean surface of about 70 centimeters (28 inches), followed by a drop of 30 to 40 centimeters (12 to 16 inches). Other data deduced from the satellite record indicate that the wave was traveling about 404 knots, with a wavelength of about 267 miles and a period of about 37 minutes.[1]

There is some possibility that the stresses have not been fully relieved and that the northern section of the fault could move and create another tsunami. If this were to occur, it would have a greater effect on the northern portion of the Bay of Bengal than occurred during the December 26, 2004, earthquake and could lead to a huge loss of life in this densely populated region. There is also a chance that a large section of the Sumatra Trench, to the south of the original rupture zone, could break in a magnitude 8 to 9 event that could cause a devastating tsunami for the southern islands of Sumatra and the northwest tip of Australia. This section, and the section of the trench to the north of the December earthquakes, probably have been "loaded" with additional stress at their edges by the December and March events.

Due to the lack of seismic sensors and water level instruments in the Indian Ocean, the center did not know whether or not a tsunami had resulted from the December 26, 2004, earthquake. Scientists believed that a locally damaging tsunami was likely, but it was not until four hours later that they realized that an oceanwide destructive tsunami existed and additional warnings were issued for the Seychelles, Diego Garcia, and East African countries. Thanks to the warning, successful evacuations were carried out in Kenya, sparing many lives. But there was no effective system for evacuation in Somalia, and there the death toll was higher.

Plans for an Indian Ocean tsunami warning system are being debated. No resolution has been reached at present, although as this book was being written there were indications that India might take the lead in sponsoring an Indian Ocean network. The Japanese have also established a regional warning presence in the Indian Ocean. Both Japan and the Pacific Tsunami Warning Center are now issuing tsunami bulletins for the Indian Ocean.

The recent Southeast Asia tsunami with its horrific loss of life and extensive destruction has prompted a review of other areas of the world that may be at risk. Areas that have previously experienced tsunami—such as the west coast of South America, notably Chile, Peru, Ecuador, Nicaragua, and Guatemala—are known risk areas.

FUTURE RISK AREAS

New research suggests that areas that have not experienced tsunami historically may be at risk. Two areas in particular could lead to huge economic loss and widespread damage if hit by tsunami waves.

California experiences damaging earthquakes every decade or so. These are generally associated with the San Andreas Fault zone or tributary faults adjacent to it. Most of these faults are inland and there has been little offshore seismic activity in recent times. However, the Southern California coastal area is crossed with a number of deep submarine canyons between Point Conception and the Mexican border. These canyons are basically extensions of the steep canyons that can be found along the coast. Southern California residents are familiar with the frequent landslides that occur in the coastal areas along Malibu, Palos Verde, and Laguna, especially following a heavy rainfall when the soil becomes saturated and fails. In the Malibu area, virtually no winter goes by without at least one closure of the Pacific Coast Highway due to a landslide.

The slopes offshore are composed of thick layers of sedimentary materials that have built up over the ages. These are saturated and characterized by a low shear strength. Offshore mapping indicates that there are areas where sliding and slumping have already occurred.

New research indicates that a large enough earthquake—say, a magnitude 7 or greater—could cause a submarine landslide in one of the deep offshore canyons, dislodging a large amount of sedimentary material from the walls of one of these canyons. If the volume of material dislodged is sufficiently large, the effect could be similar to the Papua New Guinea (July 17, 1998) tsunami. Although the earthquake was only a magnitude 7, its proximity to the coast and the steep continental shelf caused waves up to 49 feet high and thousands of fatalities.

One place where this might happen is near San Pedro, California, offshore from Point Fermin. Wave heights could reach as much as 66 feet against the Palos Verdes Peninsula bluffs directly opposite the epicenter, decreasing to around 20 feet as the wave moves west and north toward Point Vicente and then 6 feet at Palos Verdes Point. Due to the high bluffs, little damage would occur here. To the south and east, a 13-foot-high wave would impact the Los Angeles and Long Beach harbors, and the low-lying coastal areas near Alamitos Bay, Seal Beach, Huntington Beach, and Newport Beach would be hit with waves 6 feet high.[2]

Each winter the city of Long Beach erects a sand berm along the beachfront to protect residences from the run-up of water during winter storms, indicating that this area would be susceptible to damage from even a small tsunami wave. Seal Beach is known to undergo extensive flooding in the winter, when there is a combination of several days of rainfall, low atmospheric pressure, and a winter high tide. During the winter of 2005, flooding extended inland to the Pacific Coast Highway, causing it to be closed until the waters receded.

The Newport Peninsula sits 9.7 feet above mean sea level. This means that a 10-foot-high tsunami wave would overflow the peninsula. The range of high tides is 4 to 7.2 feet, so if a 10-foot-high wave hit at higher high tide, it would be 7.5 feet (17.2 minus 9.7 feet) *above* land and the surge would probably reach the upper stories of houses along the peninsula. The water would flow into Newport Harbor; in-undate Lido, Balboa, Bay, Harbor, Linda, and Collins Islands; run up against the cliffs along the Pacific Coast Highway; and then backwash into the harbor and across the peninsula into the ocean. The older single-story structures would be demolished, and it is doubtful that any of the newer structures would survive without major damage.

Ironically, a few weeks after I wrote the above paragraph—on June 14, 2005—a magnitude 7.2 earthquake occurred 80 nautical miles off the coast of Northern California, west-southwest from Crescent City, and around 485 nautical miles from Newport. At tsunami speed, it would have taken a fraction of an hour to hit Crescent City, and a little over an hour to hit Southern California. A tsunami alert was issued for the Pacific Coast but was canceled about an hour later. Some communities had no plans to deal with the emergency. Lifeguards were dispatched to clear the beaches at Newport, and a number of people evacuated their homes. My neighbors reported traffic jams trying to get off the peninsula. I was in Washington, D.C., at the time—well out of tsunami range—and was therefore spared the ordeal. However, reading the newspaper accounts the next day, after I returned home, was chilling. It is clear that had this been the real thing, the warning would have been too late to do any good, and many people would not have received the warning or would have ignored it. Those who might have tried to leave the peninsula would have been caught in a huge traffic snarl.

The authors of the Palos Verde tsunami scenario used a detailed econometric model to study the cost of such an event to Southern California's trillion-dollar economy. The estimate included direct losses in the inundated areas, the costs of business interruption and loss of jobs, the economic effects of closure of the ports of Los Angeles and Long Beach for up to a year, and damages to transportation systems and other infrastructure. Four different outcomes, with increasing degrees of severity, were considered. The estimated economic loss was in the range of $7 billion to $40 billion.

I've personally experienced at least a dozen earthquakes in Southern California and, as a member of the Earthquake Engineering Research Institute, have been involved in the engineering investigation of several of the larger ones and have also worked professionally in the area of seismic testing and improving seismic reliability of equipment and structures. My former residence in West Los Angeles was damaged by both the 1971 magnitude 6.7 San Fernando earthquake and the 1994 magnitude 6.7 Northridge earthquake, in the latter case losing the chimney and requiring $25,000 in miscellaneous repairs. The cost of the 1994 Northridge earthquake, which caused 57 deaths, has been estimated as between $20 billion and $40 billion. If the Palos Verde tsunami scenario earthquake did occur in the postulated location, earthquake damage inland would be substantial. This cost has not been estimated; it would be in addition to the $7 billion to $40 billion cited above.

The East Coast of the United States is generally considered a low seismic risk area. However, the entire East Coast of North America—from the Florida Keys to Gander, Newfoundland—faces the open Atlantic in a great arc of densely populated shoreline that extends from latitude 25 degrees north to 50 degrees north. Eastward across the Atlantic, a little more than 100 nautical miles west of southern Morocco, lie the Canary Islands, important waypoints for the early Spanish and Portuguese navigators traveling south along the west coast of Africa. The islands—seven major islands and several smaller islets—were formed by ancient volcanoes. Draw a line due west on the Atlantic Ocean from the westernmost island of La Palma, and after traversing some 3,300 nautical miles of open ocean, you strike land at Melbourne, roughly in the center of Florida's east coast.

La Palma is a pear-shaped island with a large central volcanic caldera called Taburiente. The caldera has a diameter of 5.5 miles and a peak height, known as the Roque de los Muchachois, 7,900 feet above sea level. There is a large canyon, open to the west that leads from the caldera down to the sea. The overall height of the island is 21,320 feet above the ocean floor. The base of the island is made of material called pillow lava that is cut in places by basaltic dykes. Around 1,300 feet above sea level, there is a dividing line between the upper layer of lava and rock. The island last erupted in 1971.

The concern about La Palma centers on its unique geology and activity. It appears that there is a rift zone running from the south end of the island north to the center, along the back edge of the caldera and a spot called Cumbre Vieja.[3] Rift is a seismologist's term for a fissure or plane along which rock may most easily split. Since the island has erupted seven times in the historic past (since 1585), and recently in 1949 and 1971, it is not unreasonable to assume it will erupt at some time in the future. The danger is that if molten lava interacts with seawater penetrating the lower level of the island, the resulting steam explosion could dislodge the side of the caldera, or a volcanic eruption itself could blow away the steep side of the island, as happened in Mount St. Helens and Krakatoa. If this occurred, millions of tons of rock, sliding down nearly 2.5 miles to the bottom of the ocean, could create a tsunami. Alternatively, underwater landslides of weak material could occur.

No one can predict whether this would be limited to a local tsunami or could affect a wider area. In an extreme scenario, a wave racing west at around 380 knots would wash over Bermuda in seven hours and would reach the East Coast of the United States in around nine hours. As the wave emerged from the deep waters of the Atlantic onto the continental shelf, its height would increase dramatically, potentially towering over buildings along the waterfront from Jacksonville to Boston. While the travel time of the wave would allow some warning to be given, it is doubtful that a mass evacuation of the entire East Coast of the United States could be carried out in a matter of hours. The damage to buildings and infrastructure, as well as the loss of life, would be appalling and unprecedented if this extreme event were to occur.

We know, in the case of tsunami, how massive forces and a huge release of energy deep beneath the ocean generate waves that are modest in size as they race across the ocean, but become powerful and extremely high as they reach shallower waters. Also, storms in midocean acting on near-surface waters produce large waves. In the next chapter I want to discuss what happens when several storms occur in different locations, their waves radiating out and dispersing and at some point meeting each other. To a vessel located at the intersection of several wave paths, the seas can be chaotic, with waves of all heights seeming to approach from a bewildering mix of directions. This condition is known as a confused sea.

8

A Confused Sea

hen you are on a long passage in the open ocean, you will find that a certain rhythm is established. The motion of the vessel, repeated hundreds of times every hour, becomes a ballet that sailors sense almost without thinking. It guides their movements along the deck or within the cabin. They know instinctively when to lean to the starboard because in a moment the boat will roll to the port side and so on. This ballet of movement is accompanied by a symphony of sounds—the sound of the rigging, the creaking and groaning of mast stays, the sloshing sound of water passing the hull; all become part of a familiar melody.

My senses become finely tuned to sound and movement. At night, asleep in *Dreams* when someone else is on watch, if the roll of the boat shifts slightly, or if a certain sound changes pitch or a "clunk" becomes a "clink," I find myself instantly awake, asking myself, "What changed?"

Indeed, what changed may have been a slight shift in the wind that will be adjusted for by the wind vane that steers the boat effortlessly as we stand watch, or it may have been a subtle change in the wave patterns as *Dreams'* course intersects a new set of swells and the boat's motion establishes a new equilibrium.

At any point in time the sea condition is a reflection of every storm and disturbance that has occurred within the past few weeks. Distant storms may have churned the ocean 1,000 nautical miles away from your position; a day and a half later the first swells from that storm reach the spot where you sail. For example, sailing northwest in the direction of the oncoming swell, you are likely to encounter higher waves as time passes, until finally the size of the waves once again diminishes as the last waves of that particular storm pass under the hull of the boat.

Now suppose that at the same time, the wind begins to freshen in the northeast, signaling the arrival of a nearby storm or squall. All afternoon the wind from the northeast builds, soon reaching the point at which you reduce sail to just the mainsail with one reef. Soon you are sailing at 7 knots, making good progress to a distant port; meanwhile the squall has pretty much blown itself out. What of the seas at this point?

Looking at the waves, it is difficult to discern a single pattern or uniformity. Instead, superimposed on the swell coming from the northwest are smaller waves from the local storm. Although the wind was out of the northeast, and thus at a right angle to the direction in which you were sailing, the storm waves do not line up downwind. Instead, wave trains arrive at various angles to the direction of the wind—some piling up on top of the swell, some opposing it, and some hitting it at an angle. The combined effect is to give the sea a chaotic appearance—any singular pattern of wave motion is difficult to perceive.

Being close to the local storm, dispersion had little effect and the waves reaching the boat had many different wavelengths. Another way of stating this is that the "sea" has a wide bandwidth spectrum (meaning waves of many different frequencies) compared to swell, which has a narrow bandwidth (because of the filtering effect of dispersion).

You might ask, "What is the significant wave height under these conditions?" Trying to determine how high the waves are in chaotic seas might seem to be nearly impossible, but weather forecasters have devised methods, based on the randomness of chaotic seas, that they use.[1]

The chaotic nature of actual seas is also influenced by the vicinity of coasts or islands or by areas where the ocean depth changes significantly, such as near a continental shelf. Here additional phenomena cause wave interactions.

As waves approach the shore, they slow down in shallow water. If the wavelength is long relative to the depth, the front crest (in shallower water) will travel at a slower speed than the distant following crest (remember that in shallow water, wave speed depends on depth). In shallow water, the crests also become higher. When a wave train reaches the coast or an island, a portion of the energy is lost as the waves break and a portion is *reflected* back in the form of new waves. Other phenomena can change the direction of the waves, causing them to meet and interact with the incoming waves. This interaction, which is discussed in Chapter 3, can result in constructive or destructive interference and inevitably produces more complex wave patterns.

WAVE INTERACTIONS

Waves can be *reflected*, *refracted*, or *diffracted*. Reflection, refraction, and diffraction are common to other energy forms exhibiting wavelike behavior, including light and sound. In the open ocean, which is constantly traversed by long-wavelength waves, islands and continents are barriers that reflect and refract waves. In the absence of storms or other unusual conditions, the wave patterns near these obstacles remain relatively constant throughout the seasons. To visualize this pattern, picture a small boulder in the middle of a rapidly flowing brook. The stream flow is both reflected from the upstream side of the obstacle and refracted around it. Since the water close to the boulder travels a longer distance, it has to speed up in comparison to the water closer to the bank. Downstream from the rock, the currents merge and once again move uniformly. Immediately downstream from the boulder is a spot of calm water, where the current eddies past the boulder. If the flow is smooth enough, various wave patterns can be seen radiating from the boulder.

Now shift your imagination to the vast expanses of the ocean and picture an island as an obstruction to the prevailing currents and swell.

In an analogous manner, the patterns of waves reflecting and refracting from the island can be observed. For example, as I sail *Dreams* northwest around the east end of Catalina Island, I encounter the predominant west swell "on the nose" as sailors put it, where it refracts around the end of the island. Close in, in the lee of the east end of the island, there is little wave activity, while farther out there are confused interacting wave crests as the wave trains that have traveled down each side of the island cross over each other. On the west end of the island, the incident westerly swell surges against the island cliffs in some places, breaks against rocky shores in others, and then reflects back into the oncoming waves and creates a choppy and confused sea.

REFLECTION

Consider the simplest case first. Waves approach a shear, straight cliff that drops deep into the ocean. Since the water at the face of the cliff is very deep, the waves do not break; they surge up against it and are reflected back. If the cliff is long and straight, and the waves approach in a perpendicular direction, a pattern of standing waves could be set up. However, this is unlikely; it is more realistic to assume that the waves strike the cliff at some angle and are reflected back at an equal but opposite angle. Now consider the other extreme: a sloping beach. Here waves approach the beach, slowing down as the water becomes shallower, the wavelength diminishing, and the wave crest becoming higher and higher until the wave either becomes unstable and breaks or surges up on the beach and reflects back into the sea. If the wave breaks, much of its energy is expended and reflection may be small or nonexistent.

REFRACTION

Refraction is the process by which light or other energy forms are deflected by passing through the interface of two media having different densities or by passing through a medium with varying density. A simple example with which we are all familiar is a stick held in a pool of clear water (imagine trying to spear a fish). Here the media are air and water; as you look into the water the stick appears to bend because light is refracted at the interface between the air and the denser water medium. This is because the speed of light in air is greater than in water. In an analogous manner, ocean waves are refracted as the ocean depth shoals (becomes shallower) and the waves slow down as explained previously. This explains a common phenomenon that puzzled me until I began studying waves.

The coast of Southern California takes the shape of a large arc, curving southeasterly from Point Conception (a short distance north of Santa Barbara) to the Mexican border. Along the coast are numerous bays and other irregular features, including beaches that face in directions ranging from northwest to west or south. As mentioned above, the predominant swell is from the west or northwest—although storms in Mexico occasionally will create a strong southerly swell. Yet go to any of the beaches along the coast and the direction of the waves as they come rolling in to break on the beach is *perpendicular* to shore. How can this be?

As waves approach the shore at an angle, that part of the wave closest to shore will "feel" the shallow bottom first and begin to slow, so its angle of incidence changes. As the rest of the wave gradually passes that same depth, it too slows. This process repeats again and again as the water becomes shallower and shallower, effectively altering the wave direction until the waves are perpendicular to the shore and once again are all moving at the same speed and in the same direction.

Coastlines on the mainland or on islands are more often irregular, featuring coves and bays or capes and headlands jutting into the sea. Consider first a canyon that can be seen onshore and extends out into the sea, forming a bay at the coast and an underwater canyon or deeper area immediately offshore. In this case, refraction causes the incoming waves to spread out, or diverge, and the wave intensity into the bay is reduced. It is the opposite situation in the case of a cape or point. Here, if the land projects out into the ocean and the sea is shallower than the surrounding area, the waves will be focused on the cape or point and are said to converge at that location. This helps explain why ships seek safety in bays but run aground on capes and points during storms.

Refraction is an important consideration in harbor design. Depending on bottom conditions and the geometry of the harbor entrance, refraction can cause waves to dissipate energy and be smaller in the harbor. Or, under certain conditions, refraction can focus wave energy into the harbor, creating larger waves. Through refraction, the narrow opening of a harbor can concentrate waves within a confined space. For this reason, understanding the prevailing local conditions of wind direction, current, wave periods, and wavelengths is extremely important to harbor design. With good design and the right dimensions, a harbor can be designed to attenuate incoming waves and produce calm waters. If prevailing conditions are not understood and considered in harbor design, exactly the opposite can occur; waves inside the harbor are larger than those outside. As another example, refraction causes waves to focus and become higher as they approach a submarine ridge, but to defocus and spread over a submarine valley.[2]

For similar reasons, entering any harbor can be tricky when big seas are running. Storms can alter conditions at harbor entrances in a matter of days or sometimes hours, moving sand and creating shoals—and atypical waves can toss an unwary vessel sideways.

DIFFRACTION

Diffraction is the spreading or focusing of waves by a narrow aperture or the edge of an obstacle. If waves impinge on a narrow channel such as the mouth of a harbor or a breakwater, diffraction occurs. Diffraction can cause some wave energy to propagate in a direction perpendicular to the direction of the waves.

For example, a cape or point jutting out from the coast will diffract the incident waves, causing them to change direction and "bend" around the point, as shown in Figure 17. This sketch is a view of the entrance to Newport Harbor as I observed it one morning from a plane taking off from the Orange County Airport. The prevailing westerly swell can be seen approaching from the right. As it impacts on the west breakwater, it bends to the northeast, in the direction of Little Corona Cove. Some of the diffracted waves bend around and roll into the main channel between the breakwaters.

A similar effect takes place when waves hit a gap in a harbor break-

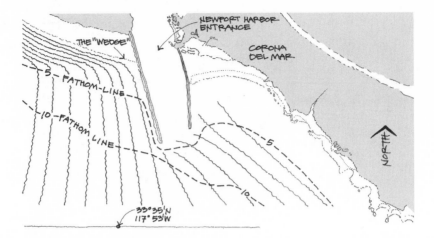

FIGURE 17 Wave diffraction, Newport Harbor entrance.

water. Within the harbor, the incoming waves are diffracted to both sides of the gap. These waves in turn can interfere with both incoming and reflected waves, producing very complex wave patterns. It is possible that under certain conditions, substantial wave action can take place within the harbor behind the breakwater.

Even though making landfall in rough seas can be hazardous, that is not the main focus of this chapter. As storms arise in distant parts of the ocean, their waves disperse and eventually impact landmasses or islands. The reflected or refracted waves from these distant storms create complex wave patterns, often superimposed on the dominant swell for that area.

POLYNESIAN NAVIGATION BY WAVE PATTERNS

The early Polynesian navigators discovered that there were repetitive patterns in the confused seas they experienced as they ventured from island to island. They saw that swell in the Pacific followed predictable paths, depending on the time of year. When one of the prevailing swells struck an island, certain wave patterns were established in a manner analogous to the boulder in the center of a rapidly moving stream described above.

The navigational capabilities of Polynesian sailors became apparent at the time of Captain James Cook's first voyage. In his July 13, 1769, entry in his sea journal, Cook relates that he was able to convince a priest and navigator named Tupia to accompany him on the continuation of his voyage when he sailed away from Matavai Bay, Tahiti.[3] From his experiences with Tupia, Cook concluded that the Polynesians were fully capable of sailing from one island to another, for a distance of several hundred leagues (more than 600 nautical miles.) Over a period of time, Polynesians populated a vast area of the Pacific Ocean, more than 4,800 nautical miles long and 3,600 nautical miles wide. In most cases, they did it on island-hopping journeys of 50 to 200 nautical miles.

How did they do it? How did they manage to avoid deadly reefs and make accurate landfalls on remote islands without benefit of compass, chart, or sextant?

In the 1900s it became apparent that some navigators who knew the ancient techniques were still alive, but they were the last of a small group that was slowly dying off. In 1965, while circumnavigating the world in a 40-foot catamaran, sailor David Lewis learned the rudiments of their techniques and then successfully sailed from Tahiti to New Zealand, a distance of around 2,200 nautical miles, without using navigation instruments.[4]

About 15 years later, after conducting some preliminary research, a young sailor, Steve Thomas, took passage to Satawal Island in the Caroline Island group, where he succeeded in apprenticing himself to a navigator named Mau Piailug. Thomas lived as a member of Piailug's family, learned the navigational techniques that had been handed down orally from generation to generation, and documented them.

Modern navigation uses the compass to establish heading (direction of travel), charts to locate the position of the vessel relative to its destination, and instruments such as chronometers, sextants, loran, or global positioning satellites to "fix" (determine) the position of the vessel.

In the Polynesian system, an initial heading was established by back-sighting landmarks on the departure island; from this, an estimate of the set of the current could be made and the course adjusted appropriately. By lengthy training and memorization, the navigator learned the positions of 32 prominent rising or setting stars, arranged in a manner analogous to the points of a compass. During the night the vessel was steered in the direction of a star known to rise over the destination island. Once this star had risen too high above the horizon to be useful for steering, the navigator steered to a second star that rose in the same direction, or to an alternate star that was "offcourse" a known amount, for which the heading of the vessel was adjusted.

The rising and setting stars constitute a type of star map. Thus Polaris marks north, rising Little Dipper (*Ursae Minoris*) marks roughly 15 degrees; rising Big Dipper (*Ursae Majoris*), 27 degrees; rising Vega, 40 degrees. Similarly, setting Little Dipper marks roughly 345 degrees; setting Big Dipper, 333 degrees; setting Vega, 320 degrees. Other stars fill out the remaining points on the star map; the formal name for such a system is a *sidereal compass*.[5]

The Polynesian navigation method is unique in another respect. The navigator imagines that his vessel is fixed and the destination island "moves" into position under various stars, until the island reaches a position under the star that tells him he has arrived. Contrast this to today's navigation method whereby a vessel is moving and the destination island is a fixed point toward which the vessel moves.

During daylight, when no stars were visible, the navigator made use of the ocean swell to judge the heading of the vessel. Navigators were trained to recognize eight dominant swells. During the winter months, dominant swells came from the northeast or east under the influence of the trade winds. With a northerly heading and an east swell, the navigator knew the heading was correct if the vessel rocked. On an easterly course into an east swell, the vessel would pitch, bow rising and then falling as the vessel rode over the swells. To maintain a northwesterly heading, the navigator would adjust the sail and steering oar positions until the vessel responded with a combined pitching and rolling motion in the right proportions. The best way to do this, I've been told, is to lie flat on your back on the deck and look up at a cloud or star. The motion of the boat is easier to sense this way than staring at the horizon.

To pass on the knowledge from generation to generation, navigators constructed "charts" made of sticks laced together to illustrate the wave patterns, using shells to mark the positions of islands.[6] Examples can be seen in the Bishop Museum, Honolulu, and in the Newport Harbor Nautical Museum, Newport Beach, California. The most skilled navigators could sense when they approached their destination by subtle shifts in the motion of their vessels as the wave pattern changed. They could recognize a number of different wave patterns; each was given a specific name. Considerable practice and skill are required to navigate in this manner. In addition, certain seagoing birds, known to frequent specific islands, would start appearing at distances of 20 to 30 nautical miles. Floating vegetation and even certain species of fish could also provide positive identification.

Once, while on a fishing trip out from Midway Island, we left in the morning at dawn. Later in the day I happened to look back in the direction of the island. The low-lying island was now over the horizon and invisible, but its location was clearly evident by a turquoise-green color in the clouds above it. This was caused by sunlight reflecting the island and lagoon onto the clouds and could be seen at a great distance. Clouds are formed by moisture-laden warm air rising over islands. At other times, a bright column or glow can be seen on the horizon, due to the reflection of the sun or moon from shallow water or a lagoon; this is known as the *loom* of land.

Because the Polynesians had no means of determining longitude, on longer voyages they usually sailed north or south to the approximate latitude of their destination. They watched *zenith* stars, or stars known to pass directly over the island to which they were steering. Once they came abreast of that position, they would tack and run downwind to their destination.

In the 1970s, a group of Hawaiian researchers constructed the *Hokule'a*, a 60-foot-long replica of an ancient Polynesian double voyaging canoe. *Hokule'a* departed from Honolua, Maui, on May 1, 1976, and reached Papeete, Tahiti, 33 days later. The navigator on that trip was Mau Piailug, whom Stephen Thomas was later to meet and study under. Also on the crew were Ben Finney and Tommy Holmes, two of the cofounders of the Polynesian Voyaging Society; David Lewis, who had studied Polynesian navigation methods in the late 1960s; and 11 other crew members. Under Piailug's guidance, the 3,000-nautical-mile trip was accomplished without instruments. The voyage was significant because it established beyond question that the Polynesians, using the ancient methods of navigation, were able to explore and eventually settle vast expanses of the Pacific Ocean.[7]

After *Hokule'a* returned to Hawaii, there was strong interest in further explorations, and in 1978 *Hokule'a* set sail again. However, the vessel encountered rough weather almost immediately, and before leaving Hawaiian waters, a tragedy occurred. Through my brother, Ken Smith, a well-known water polo coach and educator in Hawaii, I was introduced to Marion Lyman-Mersereau, who told me about her experiences on *Hokule'a*.

Marion: "We left near sundown on March 16, 1978, from Oahu. Our captain wanted to delay the departure because of the weather—heavy wind and seas—but was overruled. *Hokule'a* had rough seas crossing the Kaiwi Channel and by midnight was past Penguin Bank, somewhere off of Molokai, between Oahu and Lanai, on a southeast heading. Winds were gale force (35 to 40 knots) out of the northeast, impacting the port side of *Hokule'a*. Swells were running 6 to 8 feet. We became aware that the starboard hull was taking on water and *Hokule'a* was riding low on the starboard side. The captain rousted everyone to sit on the port hull to help balance the canoe. We also came off the wind so the crew could shorten sail. Before we could get the sail down we were hit by gusts of wind and a large wave and *Hokule'a* capsized. Besides me, the crew included our navigator, Nainoa Thompson, Eddie Aikau, and 13 others. I was the only woman. When we capsized, we scrambled back to hang on the upside-down hulls, consoling ourselves by saying that as soon as it became light someone would surely see us.

"By the next day we were drifting farther away from the islands—both Molokai and Lanai could be seen in the distance—but no rescuers appeared. Several of the crew—including me—were sick and everyone was wet and tired. Later that morning Eddie volunteered to go for help on a surfboard. The officers gave their approval, as with his reputation and skill it seemed like our best hope at the time. He was asked to take a lifejacket; but later this came floating back, so no one knows if he ditched it intentionally or it was lost when something happened to him.

"Eddie was never seen again after he paddled away from us that morning. It was a terrible tragedy. Everybody liked Eddie. He was well known, not just in Hawaii, but internationally, because of his surfing abilities. But he never lost his lifeguard instincts, from all those years of being a lifeguard at Waimea. He wanted to get help for us before the canoe was lost. As it was, we were in the water for 22 hours. Eventually *Hokule'a* was spotted by a passing Hawaiian Airlines pilot and coast guard helicopters picked us up that night. When the coast guard showed up we told them about Eddie and they launched an intense search for him. Some of the crew stayed with the *Hokule'a* until it could be towed back. It was salvaged and repaired and made many more trips for thousands of miles through Polynesia and Micronesia."

I happened to be in San Francisco in 1995 as part of a large welcoming crowd when *Hokule'a* sailed down from Seattle and passed under the Golden Gate Bridge. It was a very moving experience. By then, *Hokule'a* had sailed the equivalent of halfway around the world on its numerous trips to Polynesia.

On several trips to Catalina Island in *Dreams* I charted the predominant swell that comes from the west and how it is modified by reflection, refraction, and diffraction around the island. From numerous trips I was aware of the swell but previously never paid attention to the subtler aspects of how the boat moved under its influence. Hearing about *Hokule'a* and Piailug, and listening to Lyman-Mersereau, kindled my interest. On the next trip I made to Avalon, Catalina Island, I closed the compass binnacle and tried to steer by keeping the swell at 45 degrees on my starboard bow. (The heading for Avalon from Newport is 225 degrees magnetic, so for a west swell I needed to subtract 45 degrees.) The island was obscured by early morning overcast skies. When it finally cleared a few hours later, I could see I was on course.

Coming back was a different story. By then a south swell had set in and we had confused seas. As I looked astern, it seemed the waves came first from one direction and then another, and sometimes I could not determine from what direction they were coming! I could see that it would take considerable practice, along with the ability to sense the predominant swell, to become proficient at this method, especially over much longer distances. I have great respect for the Polynesian navigators thanks to my amateur experiments.

SOLITARY WAVES

A somewhat unusual phenomenon is the solitary wave that travels on the surface or beneath the surface of the sea. Such waves are called *solitons*. Originally, as in the case with other unusual marine phenomena, the existence of such waves was doubted. The advent of satellite observation of the oceans has provided evidence of their existence. As many as seven solitons have been captured by satellite photographs crossing the Andaman Sea at the same time.

Today, because of the December 26, 2004, Andaman-Sumatra earthquake and tsunami, more people are aware of the Andaman Sea than before. It is a relatively shallow body of water lying east of the Andaman and Nicobar Islands and west of the Malay Peninsula. Theory has it that tidal currents squeezed between the islands generate internal (subsurface) solitons. As these propagate, they interact with the surface of the sea to produce a large number of randomly directed small waves, making the sea surface look as if it is boiling. The "boiling" sea appearance has also been photographed by a drill ship working in the area. On a much larger scale, on February 17, 1846, the harbor on the island of Saint Helena was suddenly violently agitated, causing the loss of 13 vessels. At distances of more than about 1,600 feet from shore, the water was calm.

No one is sure how solitons arise in the open ocean. It may be that long-wavelength (faster-moving) waves overtake slower waves, canceling out a portion of them and creating a single wave group that is compact, higher, and long-lived. This resembles a pulse that approximates and propagates as a solitary wave.[8]

CONFUSED SEAS

It is not uncommon to have more than one storm at a time in different parts of a sea. Also, any given storm is most likely moving. The net result of this is that multiple wave trains can be crossing the sea at a given location, intersecting at various angles with waves from other storms or with waves reflected or refracted from some distant shore. As mentioned earlier, these waves can interfere with each other. If the wavelengths are nearly equal and the waves are in phase or nearly in phase, they can combine and create a larger wave. Or, if the wavelength and phasing are just right, one wave can tend to cancel another. If they intersect at an angle, complex diamond-shaped wave patterns can be produced. Wave patterns lose all semblance of regularity, and instead become random and irregular. Under these circumstances with severe storm conditions and large waves, the seas seem to come at a boat from every angle, creating a chaotic situation as the helmsman tries to maneuver the boat. This condition is described (with a certain degree of understatement) as *confused seas*. Navigating in confused seas can be particularly harrowing in rough weather.

Confused seas, combined with a rogue wave, created one of the most bizarre transatlantic passenger liner incidents I encountered in my research. The vessel was the SS *Pennsylvania*; she was barely a year old when the incident in question took place. The SS *Pennsylvania* was a football field long, at 336 feet, with a beam of 43 feet. She was equipped to carry both passengers and freight, and could accommodate 76 first-class passengers and more than 800 persons in "steerage class."

The period commencing with the winter of 1873 and continuing until the winter of 1874 saw a number of severe storms roil the North Atlantic steamship routes, causing the loss of 12 passenger ships. One of them, a White Star Line vessel appropriately named the *Atlantic*, went down with a loss of 585 lives on April Fools Day, 1873. One of the survivors was a man named Cornelius Brady, the third officer on the *Atlantic*.[9]

As the SS *Pennsylvania* prepared to sail on February 21, 1874, from Liverpool to Philadelphia, she carried only 12 steerage passengers and one first-class passenger. However, at the last minute a second first-class passenger boarded the vessel. Fortuitously, the last passenger happened to be Cornelius Brady.

The SS *Pennsylvania* entered the Irish Sea, traveled southwest past the east coast of Ireland, through St. George's Channel, and into the North Atlantic. After several days of steaming she encountered increasingly foul weather, culminating in a terrific hurricane that pounded the ship with heavy waves and confused seas. Heavy waves broke over the bow, damaging the front portion of the vessel and flooding the lower decks. Around midnight, a mountainous rogue wave swept over the vessel, demolishing the bridge, crushing the forward hatches, ripping loose and carrying away lifeboats, life rafts, much of the vessel's superstructure, lifelines, stanchions, and rails—and the crew navigating the boat!

All of those who had been on the bridge fighting the storm—the captain, first mate, second mate, and two crew—were lost overboard. But this fact was not known to the passengers and remainder of the crew, who were fighting for their lives belowdecks by trying to stop the flooding and by reinforcing the damaged portions of the vessel. The third officer—apparently paralyzed by fear—cringed belowdecks and would do nothing. At this point, Brady and a group of passengers and crew took it upon themselves to replace the damaged hatches with others from belowdecks to try to halt the influx of seawater. Brady sought the approval of the captain for this measure; at this time the loss of the bridge and all of the ship's officers who had been on deck was discovered. Brady went on deck, saw the extent of the damage, and directed the crew to slow the vessel and heave to in a better position to ride out the storm. Brady secured a vote of confidence from the surviving crew members and the passengers, took charge of the vessel, and succeeded in bringing the SS *Pennsylvania* safely into Philadelphia—albeit six days late—where he was accorded a hero's welcome.

SURVIVAL STRATEGIES

Under stormy conditions the helmsman's goal is to keep enough way (forward movement) on the vessel so that a safe course can be maintained. (At very low speeds, the rudder has little effect, and steering the vessel becomes almost impossible.) There are several options, assuming there are no nearby rocky shores. The first is to "run" before the storm, or keep the wind and waves behind the vessel and go in the same direction as the waves. If this is not possible, the next choice is to slow down and head into the waves, taking them on the bow but at an angle to minimize slamming of the vessel as it rides up and over the wave. The third alternative is to "heave to," which means to position the vessel so it remains nearly stationary in the water, with a slight sidewise motion.[10]

With high winds and heavy seas, it is important to reduce the vessel's speed and to maintain a safe course relative to the wave direction. If the speed is too high, there is a risk that the vessel will sail rapidly over the crest of a wave, bury its bow in the trough, and then flip end over end. This is known as *pitchpoling* and invariably results in damage to the vessel and its occupant, and possibly loss of the vessel.

If a vessel is traveling too fast—even after reducing sails to a minimum—it becomes necessary to slow it by other means, including dragging long lines, called *warps,* behind the vessel (or in front of it if the goal is to keep the bow headed into the waves) or deploying a drogue (small parachute). These methods are collectively called *sea anchors* and are used to slow the vessel in high winds. In addition to slowing the vessel, they help the helmsman keep the vessel on the right heading so a large wave does not break broadside on the vessel and roll it over. Sea anchors work for yachts and smaller ships, but are impractical for commercial vessels. Any vessel is designed to be self-righting through a specific angle. A weighted keel or ballast in the hold of a vessel brings it back to a vertical position. However, once a critical angle of roll is exceeded, the restoring moment of the keel is ineffective, and the vessel can roll over completely.

Under heavy seas the helmsman typically wants to maintain a course down seas at a slight angle to the waves. This way the vessel will ride up on the oncoming wave and slide down the crest (rather than nosedive straight down into the trough). The helmsman's objective is to keep the big waves astern and not let the vessel turn sideways to the waves, resulting in *broaching* or rolling over.

When the seas consist of swell from several distant storms, each wave train will, through dispersion, separate into a group of waves with a predominant period. When these waves come together, they will combine in a pattern that is determined by their respective wavelengths. The resultant composite wave will have a series of low crests where the waves interfered destructively, followed by several large ones where they did so constructively. These conditions create the wave "sets" familiar to surfers. With practice, the helmsman also can determine the rhythm of the waves, steering the desired course when the waves are smaller, and falling off the course when the large waves come, to avoid unnecessary pounding of the vessel.

The helmsman of a vessel faces many hazards in heavy weather. In confused seas, life is much more difficult, because the vessel is likely to be battered and tossed from multiple directions, from time to time being buried under tons of water landing on the bow or in the cockpit. Under these conditions, the sturdiness of the vessel and how water-tight it is (windows and hatches) are as vital as the seamanship of the captain. Constant vigilance is required. Staying with the ship is usually the best idea, although not all would agree. One man in recent times carried the idea of "staying with the ship" to a remarkable extreme.

STAYING THE COURSE

Late in December 1951, the American cargo ship *Flying Enterprise* sailed from Hamburg, Germany, for New York. The *Flying Enterprise* was about 400 feet long and carried a cargo of pig iron, coffee, and other goods, as well as 10 passengers. Her master was Captain Henrik Kurt Carlsen, with more than 20 years of sea experience, including 40 Atlantic crossings. By Christmas day, the *Flying Enterprise* had passed through the North Sea, gone southwest down the English Channel, and entered the North Atlantic—just in time for one of the worst storms of the decade.[11]

The storm extended from Norway at the latitude of the Arctic Circle to Gibraltar in the south. A number of ships were lost in hurricane-force winds and seas that reached 40 to 60 feet, including the Norwegian tanker *Osthav* off the coast of Spain, the German ship *Irene Oldendorff* in the North Sea, and several others off the English coast.

By the day after Christmas, the *Flying Enterprise* had progressed to a point about 300 nautical miles southwest of Ireland. At this point the weather was so threatening that Captain Carlsen decided to heave to and go with the storm until conditions improved. Early on the morning of December 27, *Flying Enterprise* was hit on the starboard side by a rogue wave more than 60 feet high. As the ship rolled to port under the impact, cargo shifted, the crew reported hearing the cracking sound of rending steel, and the vessel took on a heavy list to port. When the crew inspected the vessel, they found water flooding the number three hold and two cracks extending across the weather deck. The crew attempted to make emergency repairs but later that day the engines stopped and could not be restarted. Huge waves continued to batter the vessel and the *Flying Enterprise* listed 45 degrees.

On December 28, Carlsen radioed for assistance; half a dozen vessels answered the SOS. However, it would be another day before they could reach the stricken ship. On December 29, as a number of vessels were standing by, he gave the order to abandon ship. The starboard lifeboat had been smashed to kindling by the rogue wave and the port lifeboat was unusable due to the list of the vessel, now around 50 degrees, so the only alternative was for passengers and crew to jump into the water, where they were picked up by lifeboats from the waiting vessels including the SS *Southland* and USNS *General A. W. Greely*.

Captain Carlsen refused to leave the *Flying Enterprise*. The crew and passengers safely off and his vessel still afloat, he was determined to see her into port so the cargo could be salvaged. On December 30, he again radioed for assistance, this time seeking a seagoing tug to tow the *Flying Enterprise* to the nearest port. Thus began Carlsen's incredible one-man fight against the sea, a modern Danish saga that would continue for another 12 days.

Imagine trying to survive on a ship that constantly is being battered by waves, capable of sinking at any moment, with everything wet. Just trying to walk around with the decks at an angle of 50 degrees was a challenge. Half the time he'd be walking on the walls and bulkheads of the vessel! This was how Captain Carlsen welcomed in the New Year, still refusing entreaties from several ships now on the scene that he abandon the *Flying Enterprise*. "Not while she floats," was his implicit answer.

On January 3, 1952, the British deep-sea salvage tug *Turmoil* finally reached the *Flying Enterprise*. For two days, Carlsen tried to connect a towline, but in the rough seas, single-handedly, he was unable to do so. On January 5, the seas had calmed considerably, and Keith Dancy, the mate from *Turmoil*, had the tug brought close to the *Flying Enterprise* so he could make a daring leap onto its deck and help Carlsen attach the towline.

With the towline finally attached and the seas moderate, the tug was able to make 3 knots toward the coast of England. Captain Carlsen finally had reason to be optimistic. The storm had blown the *Flying Enterprise* 200 nautical miles closer to England, and he now hoped that he could get the *Flying Enterprise* into Falmouth for salvage.

For three more days they progressed—Carlsen and Dancy alone on the *Flying Enterprise*, the tug in front, the USS *Willard Keith* (DD775) now on the scene as an escort vessel. On January 7, the list was 70 degrees. On January 8, only 60 nautical miles from Falmouth, the weather once again deteriorated and *Turmoil* and *Flying Enterprise* heaved to in gale-force winds. *Flying Enterprise* was rolling up to 80 degrees. On January 9, the towline separated, but the weather continued to move the *Flying Enterprise* in an easterly direction. On January 10, only 40 miles from Falmouth, the two men on board observed unmistakable signs that the ship was going down. They walked along the ship's funnel (now nearly horizontal), and stepped off into the sea—Carlsen, of course, being the last to leave. They were picked up within minutes by *Turmoil*, in time to watch the *Flying Enterprise* sink stern first into the sea.

On January 11, when the *Turmoil* docked in Falmouth, Captain Carlsen was greeted as a hero by a crowd of thousands. He received numerous honors and awards, including a ticker-tape parade up Broadway when he returned to New York. But amid the attention, he remained modest and declined to capitalize on his accomplishment, stating that he didn't think he was entitled to any special treatment, because "I failed to bring my ship back to port."

In truly bad weather, there are times when all you can do is hunker down and wait for the weather to abate. Tania Aebi sailed away from New York at age 18 in an 26-foot-long sloop to return two and a half years later in 1987 as the youngest woman ever to sail around the world single-handedly. On the last leg, after leaving Gibraltar to cross the Atlantic for home, her log book recorded her fear when she was hit by a sudden gale. Mountainous waves taller than the length of her boat engulfed her vessel from every angle, rolling it, tossing it forward, water breaking into, over, and all around the boat. On deck in the darkness as she struggled to reduce sail, she saw the silhouettes of gigantic waves bearing down on the boat like "freight trains." Unable to watch any longer, she resigned herself to her fate and went below. To get her mind off the situation, she played solitaire while the boat basically sailed itself through rough and confused seas. Forty-eight hours later— her boat still afloat—the storm finally let up.[12]

Tania was fortunate that her small boat survived this onslaught of waves from every direction. The next chapter shows that even the mightiest ships are not immune to the danger of waves so large that they are best called *extreme waves*.

9

Freaks, Rogues, and Giants

On March 2, 2001, the cruise ship *Caledonian Star*, heading home from a cruise in Antarctic waters, found itself in gale-force winds. The *Caledonian Star* is a rugged ship, specially designed and strengthened for service in the rough waters and rough weather of the Southern Ocean. At 5:30 P.M., the first officer, standing watch on the bridge, saw something totally out of the ordinary—something beyond anything he'd ever experienced before. In a BBC broadcast a year after the event, he was quoted as saying that out of nowhere a wave appeared—a wave that seemed to be twice as high as the other waves the vessel was experiencing. Later estimates, based on the height of the ship's bridge above the waterline, indicated that the wave was 98 feet high. The wave approached the vessel at an angle of 30 degrees on the starboard bow, preceded by a deep trough.

The *Caledonian Star* nosedived into the trough and tilted to one side, throwing passengers against the bulkheads as the bow of the vessel buried itself in a mountain of green water. The helmsman was unable to see the top of the wave from his position on the bridge. Both he and the first officer stated that they dove for cover as a wall of water crashed through the windows and washed them away from the ship's controls. They described being completely submerged under water; tossed about in a maelstrom of cushions, books, and debris; and literally swimming back to seize control of the ship. The saltwater wreaked havoc with the ship's instrumentation, wiping out radar, gyro compasses, some of the radio communication equipment, the depth sounders, and sonar.

Fortunately, the engines remained in operation and the steering was not affected, so the crew was able to regain control of the vessel and keep it from turning parallel to the seas where it might have rolled and suffered more serious damage. Crewmen were able to board up the broken windows, and the vessel returned to port without further incident.[1]

CAUSES AND CHARACTERISTICS OF ROGUE WAVES

The earliest seafarers experienced huge waves that swamped their frail vessels, in most cases leaving no survivors. The numerous wrecks in the Mediterranean, their hulks replete with wine bottles and other cargo, are mute testimony to these disasters. The story of Jason and his quest for the Golden Fleece predates the Trojan War and would have been known to Homer and even to Odysseus had he been an actual person. When Apollonius Rhodius rewrote the ancient story in 300 B.C., he was just retelling what the ancient Greeks had known for centuries—that the sea could be very dangerous. Only by invoking the favor of the gods, or by skillful piloting, did the Argonauts survive, as this passage relates:

> ... [They] set off under sail through the eddying Bosporus. There a monstrous wave rears up in front of their ship like a soaring mountain; it threatens to crush down as it towers high over the clouds. You would imagine that there is no escape from a miserable fate, as the violent wave hangs like a cloud over the middle of the ship. ... [2]

The language here is telling. Note the key words: "like a soaring mountain," "towers high over the clouds," "hangs like a cloud over the middle of the ship." Compare this to Captain Warwick's description later in this chapter of what happened to *Queen Elizabeth 2*, as he told his story nearly 4,000 years after the Argonaut's tale. Some of the "mysteries" of the sea are not so mysterious; we've just chosen to ignore them. Until recent times, many such reports were dismissed as the products of overly active imaginations or as exaggerated reports of actual phenomena. New research suggests otherwise.

What are rogue waves like? Until recently, there was no generally accepted definition of what constitutes an extreme wave—as noted by the fact that oceanic literature refers to them variously as freaks, rogues, or giants. Today, a definition of an extreme wave is one that is 2.2 to 2.4 times higher than the significant wave height, although some scientists might say that an extreme wave is *any* unusually large wave.[3] Extreme waves may occur singly or in groups of several waves. They differ from ordinary waves in several other ways, too. They are asymmetrical, resembling the profile of a wave that is breaking or about to break, and may be preceded or followed by a deep trough.

This description illustrates one of the important features of extreme waves: the crest height is much greater than the depth to the trough of the wave. In the mathematician's vernacular, such waves are nonlinear. Moreover, extreme waves are frequently much steeper—say, on the order of two or three times—than the prevailing waves. These characteristics—a steep wall of water rising several times higher than the usual waves, preceded or followed by a deep trough—are among the reasons such waves are dangerous to supercarriers as long as two football fields and are likely to be disastrous to fishing boats and yachts. Their occurrence is so far largely unpredictable, and they can arise suddenly, even in moderate seas. In some cases we have a good idea about the conditions that can give rise to an extreme wave—an example being interaction of swell and strong currents—but in other instances researchers are not sure. The sea state and duration are believed to be important factors.

Even worse than one extreme wave is several. This condition is called "the Three Sisters," since there have been a number of cases in which three successive extreme waves have been reported.

Any sailor who has sailed the ocean has most likely experienced a rogue wave, if you define "rogue" to signify a random wave at least 2.2 times as high as the significant waves in the sea at that time. Unless the seas were running exceptionally high and the wave presented a real danger to the vessel, such a wave was probably noted and then quickly forgotten as the voyage progressed.

I've experienced several such waves, once on the same trip to Catalina mentioned in Chapter 4, except that we were outbound for the island. There had been several storms in the Southern California area the previous week, but by the time we sailed the weather had settled down and seas were the usual 2- to 3-foot-high northwesterly swell. The trip was uneventful until we were about two-thirds of the way across the San Pedro Channel on our way to the island, when I saw a larger than usual wave—around 6 feet high. I turned *Dreams* to take it at 30 degrees. The bow rose sharply and then slammed down into the trough, sending a large spray of water into the air on both sides of the bow. Dishes and pots and pans rattled in the galley lockers. The boat shuddered, regained momentum, and we continued on our way. Someone asked, "What was that?" Oh, just a wave, I replied. I promptly forgot about the incident until I was reminded of it by Nancy Smith, while writing this book.

There are several schools of thought regarding the formation of extreme waves. Some scientists are exploring sophisticated theoretical models in an effort to better understand the underlying physics of extreme waves so that conditions that give rise to them can be predicted. At the other end of the spectrum are those who think that extreme waves simply arise as a consequence of the randomness of the sea; there are no unusual physics involved. Consequently a great deal of research is being performed to better understand how extreme waves are formed. For example, it may be that nonlinear effects, multidirectional effects (such as crossing waves), focusing, or other phenomena can cause them. See Appendix A for a brief description of these efforts.

I believe there are at least four ways in which extreme waves can be produced: (1) the interaction of strong currents with storm waves moving in the opposite direction; (2) hurricanes and storms; (3) shoaling and pile-up of waves in shallow depths along the continental shelves, the North Sea, and other shallow-water seas; and (4) seas in which constructive interference (superposition) of random waves suddenly generates a wave much higher than the average.

STRONG CURRENTS

One area where giant waves have sunk a disproportionate number of ships is the Agulhas Current along the east coast of South Africa (see Plate 12). This area has been known since ancient times as a dangerous coast and was avoided by early Arab sailors who traversed the Indian Ocean before Portuguese navigators finally rounded the Cape of Good Hope in the 1500s. Storm-produced waves originating in the Southern Ocean run up against the southbound Agulhas Current, creating extreme waves. Since 1990, at least 20 ships have been devastated by rogue waves there. Researchers say that the Agulhas waves can be 66 to 98 feet high and move at a speed of 50 knots with a wavelength of half a mile. If they topple onto a vessel that has its bow in a trough, the impact is that of thousands of tons of water. Scientists have calculated that waves there could go as high as 190 feet.[4]

In the last century, rogue waves in this area have claimed a number of major ships, some notable examples being the passenger ship *Waratah* (named for the state flower of New South Wales, Australia), the tanker *World Glory,* and the cargo ship *Neptune Sapphire.*

The case of the *Waratah* is especially unusual—one of the long-standing mysteries of the sea. She was a twin-screw vessel, 456 feet long, with a beam of 59 feet. The vessel was constructed for the Blue Anchor Line to provide passenger and freight service between London and Australia. She made this trip successfully on her maiden voyage, and was returning from Australia to England on the second voyage when she disappeared without a trace in 1909. She seemed to be seaworthy for a vessel of that period. She had 16 lifeboats capable of holding around 800 people, three rafts, more than 900 life jackets, signaling flares, and rockets. She had the usual Board of Trade and Lloyds certificates of seaworthiness.

On her second voyage, the *Waratah* made the trip to Australia successfully and on the return stopped in Durban, South Africa, where she unloaded several hundred tons of coal, then departed on July 26 for Cape Town. Around 8 or 10 hours later she was seen and hailed by a passing vessel, the *Clan MacIntyre.* As far as we know that was the last view anyone had of the *Waratah.* A number of vessels searched over thousands of square kilometers of ocean, but no sign of the vessel, her

92 passengers, or 211 crewmen was ever found. The search even extended to the Crozet and Kerguelen islands south and east of *Waratah's* expected course. The significance of these islands becomes clear if you recall Brad Van Liew's impressions of them when he passed them in the Around the World Alone Race (see Chapter 4).

At the inquiry conducted a year later, an officer of the *Clan MacIntyre* testified that by the evening of July 27, the winds were at gale force from the southwest, and by the next day had become hurricane strength. At the hearing, there was disputed testimony that the *Waratah* was top heavy and had a tendency to roll excessively. Other experts denied this. A number of witnesses claimed they'd seen the vessel sink, or produced what they claimed was debris from the ship, or even claimed they'd sailed through an area and saw bodies floating in the water. The "witnesses" were all exposed as frauds; the debris, from other vessels; the dead bodies, actually thought to be fish offal, trash, or possibly pieces of a whale. Curiously, those who "thought" they saw dead bodies floating in the sea did not bother to stop to try to recover them or even check to make sure they were bodies.

So, what can we infer now, with the hindsight of 100-plus years, not being marine architects, and with no detailed knowledge of how the *Waratah* was designed and ballasted? First, we know that when storms coming up from the Southern Ocean stir up the sea and run headlong into the Agulhas Current, there is a strong likelihood that an extreme wave will be produced. *Waratah's* last known position at latitude 31 degrees south, close to the coast, put her right where such action was to be expected—right where giant waves could build up in the shallower water. Second, there is a good possibility that the resulting waves had a wavelength of around 500 feet, the same as the length of the ship—and the precise length to suspend her in midair and break her back. Or, if she had a tendency to roll, there were no doubt waves that could roll her, and then crush and sink her. However, all of this is speculation, and we can do little more than accept the Board of Inquiry's conclusion that the *Waratah* had been "lost in a gale of exceptional violence . . . and capsized."[5]

The story does not end here. That a large vessel could disappear completely without a trace—no survivors and no witnesses—is bothersome. Human nature is such that people want answers, want to tidy up this gap in their knowledge. Consequently, between 1983 and 2000 there were 10 expeditions dedicated to locating the wreck of the *Waratah.* These expeditions were led by Emlyn Brown, a South African filmmaker and undersea explorer. The 1983 expedition failed to locate any wrecks; however, a second attempt in 1987 at a different location using side-scan sonar found the outline of a wreck in 384 feet of water. This expedition was financed by noted author Clive Cussler. Subsequent attempts in 1989, 1991, 1995, and 1997 using a diving bell, a piloted submersible, and mixed gas deep-sea diving all failed or were aborted due to weather and current. Finally, in June of 1999, improved side-scan sonar and underwater video revealed the outline of a wreck that closely matched *Waratah*'s dimensions and ship plans. In 2001 arrangements were made to bring a submersible exploration vehicle from the United States to attempt to retrieve objects from the wreck that would permit a positive identification. As the submersible vehicle approached the wreck, a strange sight appeared out of the murky water on the bottom: army tanks!

It was not the *Waratah.* Instead, the researchers had found the wreck of the *Nailsea Meadow,* a World War II cargo ship that had been torpedoed in 1942. Additional attempts were launched in 2003 and 2004, but without success. So the location of the *Waratah* remains a mystery. [6]

On June 14, 1968, the *World Glory,* a 736-foot-long tanker was positioned off the southeast coast of South Africa. While in the Agulhas Current, the vessel encountered a storm with waves building to 50 feet high. The ship slowed and headed into these seas, taking tons of green water over the bow. *World Glory* encountered a wave estimated at 70 feet high that passed under the ship, raising it so that the bow and stern were no longer supported by the sea (a condition called "hogging") and cracking the upper deck. Soon another huge wave wracked the ship, raising the bow and producing sagging forces that ripped apart the weakened section, causing the ship to break in half; both halves sank within four hours. Oil spilled into the sea and caught fire. Between 18 and 24 hours later, 10 survivors were rescued. [7]

The *Neptune Sapphire* was a cargo ship on its maiden voyage in

1973. Off the east coast of South Africa, it was hit by a single large wave. It lost its bow and 200 feet of the front half of the ship, but remained afloat and was towed to port for salvage. The salvage master remarked that it was as if a cutting torch had cut the ship in half.[8]

Other danger areas in addition to the Agulhas Current are the Gulf of Alaska, the North Sea, the Kuroshio Current (Japan), Nantucket Shoals, Hourglass Shoals (between Hispaniola and Puerto Rico), and the Gulf Stream. The Kuroshio Current is the second fastest after the Gulf Stream, flowing at 3 to 4 knots, and is known for large waves. The Gulf Stream, flowing at 4 knots, creates giant waves when it collides with high winds caused by storms coming down from the northeast.

One of the stranger rogue wave incidents I discovered in my research involved the submarine *Grouper*. *Grouper* was in the Gulf Stream midway between the Carolinas and Bermuda when it surfaced shortly before midnight to charge batteries. Several officers and crew came on deck to stand watch and breathe the fresh sea air as *Grouper* rolled in a mild Atlantic swell. The watch used 8 × 50 binoculars to scan the horizon for other vessels. It was a clear night and nothing could be seen on the horizon in any direction. In the morning, after the batteries were charged, the order was given to dive to a periscope depth of 59 feet. The deck watch climbed back down below. The last man exited the bridge, closing the hatch behind him, as the submarine began its dive to the prescribed depth. Ballast tanks were flooded to reduce buoyancy and the bow planes adjusted to angle the submarine down. Up to this point the dive was uneventful.

As *Grouper* started down, the conning officer observed the rate of dive accelerate. Suddenly he realized that *Grouper* was going past 59 feet—and going fast. The depth gauge indicated that the vessel was plunging down. He yelled to the crew, "Come on people, get me up, get me up." Crew members scrambled with the controls, blowing ballast and putting full rise on the bow planes. Still the submarine continued down, as if pushed by a giant hand, reaching 170 feet before finally starting to rise once again.

Several minutes passed. Finally the submarine started to rise, and as the periscope broke the surface of the ocean, at first nothing could be seen due to the turmoil of rough water, almost as if they were in the wake of a passing vessel. Moments later, the sea was once again in its normal state, nothing visible on the horizon.

By analyzing the rate of descent, the depth, and the known height of the submarine, the crew later concluded that *Grouper* had been overtaken by a 100-foot-high rogue wave. In all likelihood, the wave was preceded by a deep trough. As the submarine began its dive, it plunged down into the trough, its momentum carrying it down initially, and then tons of water from the crest of the wave pushed it down still farther until the apparent depth was 170 feet. At this point, the orbital motion of the wave helped push the submarine back toward the surface.

After recovering from this surprise, the cruise of *Grouper* continued uneventfully. There was no damage to the vessel. Actually, it was fortunate the submarine dove when it did. Otherwise, there might have been crew on deck when the wave hit, and they could have been injured or lost overboard.[9]

STORMS

Extreme waves can arise during gales and storms. Plate 13 shows a freighter en route from Los Angeles to Yokohama battling waves during a typhoon. A few seconds later this breaking wave poured green water as high as the bridge—60 feet above the waterline. The ship was only a few years old and with a length of 330 feet it came through the storm with no major damage. A larger vessel might have had a more difficult time of it in this storm in which the wavelength was approximately 600 feet.

When a storm system moves at the group speed of the wave, it can continuously feed energy into waves, causing them to grow. Another important cause of big waves in storms is believed to be crossing seas from two different storms; waves can add and build from the continuous input of energy, producing confused seas that batter a vessel from different angles and make it difficult to heave to or take other defensive measures. Another mechanism is when a small, fast-moving low front encounters a big low and produces extreme waves. This is now believed to be the source of the Draupner wave shown in Figure 18.

The incident involving the *Caledonian Star* at the beginning of this chapter is an example of extreme waves caused by storms. Another example is the incident of the *Bremen* while traveling in the same waters two weeks earlier.

The *Bremen* is a specially built cruise ship reinforced for service in ice-laden waters. She cruises the southern latitudes, from South America to the Falkland Islands and also to the Antarctic Peninsula. In the South Atlantic on February 22, 2001, while carrying 137 tourists, she was struck by a 98-foot-high extreme wave that smashed her bridge and knocked out all of her instruments and power, leaving the ship disabled in very rough seas for 24 hours. Unlike the *Caledonian Star*, the *Bremen* was unable to maneuver and was in a very vulnerable position. As the ship rolled broadside and was pounded by heavy seas, there was a danger of the windows of the onboard restaurant breaking out, which could have led to flooding and eventually sinking the ship. The captain must have considered what it would be like if it were necessary to abandon ship. Having passengers in open boats in these rough and icy waters would have been fatal—reminiscent of the tragedy of the *Titanic*. Fortunately, under trying conditions, the crew was able to make emergency repairs, restart the engines, and limp back to Buenos Aires and safety.[10]

Two vessels encountering giant waves within a two-week period in the Southern Ocean—surely this suggests that the occurrence of extreme waves may be more common than previously believed.

Encounters by passenger liners and cruise ships are not unusual. During the Second World War, when the original *Queen Mary* served as a troop ship, she was in the North Atlantic, south of Newfoundland, when a "freak wave" hit the ship broadside, causing her to roll until her upper decks (normally 59 feet above the waterline) were awash and the crew feared she would capsize. This would have been disastrous, because when this event occurred—in late 1942—the *Queen Mary* was carrying 15,000 American troops from the United States to Glasgow, Scotland, to aid in the defense of Britain.[11] Today, the venerable *Queen Mary*, having outraced German submarines, survived the war, and weathered many a North Atlantic storm, rests peacefully at a dock in the Port of Long Beach, California, where she serves as a floating hotel.

A few months later in 1943, the *Queen Elizabeth*, sister ship to the *Queen Mary*, experienced a similar incident in the North Atlantic south of Greenland. In this instance the ship hit a trough preceding an extreme wave. The wave came over the bow, rolled forward across the foredeck, and dumped green water on the bridge, 90 feet above the waterline. This was followed by a second wave, which smashed bridge windows and caved in an area of the foredeck 6 inches.[12] In both cases, the ships were sailing in rough seas, but the isolated single wave that hit them was much bigger than the prevailing waves.

One of the classic stories of fishing boats caught in rough seas is told in Sebastian Junger's *The Perfect Storm*. In the book is a brief reference to a boat named *Lady Alice*.[13] Captain Skip Gallimore is the owner of *Lady Alice*. Gallimore's interest in the ocean began when he was in his 20s and worked as a diver and lobsterman. He eventually bought a wooden boat and went into the marine salvage business, recovering bronze props and other salvage items from wrecks along the coast near New Jersey. In 1974 he started offshore fishing and in 1978 bought *Lady Alice*, a steel-hulled long-line fishing vessel. *Lady Alice* is 95 feet long with a 24-foot beam. In addition to the captain it has a crew of five.

In 1980, two years after buying *Lady Alice*, Gallimore was fishing the Grand Banks. He'd had a good trip and was returning to port with a hold full of swordfish. He was on the east side of the banks trying to dodge a big storm. He knew it was coming and had been monitoring its progress through the marine weather faxes. He had ordered the decks cleared, the hatches sealed, everything stored below or tied down. *Lady Alice* was holding her own by maintaining a quartering course into the seas. After 22 hours without sleep Gallimore tried to get some rest, leaving orders to be called if the weather worsened.

At dawn on Saturday, September 6, Gallimore and another crewman were in the wheelhouse. The wind had been blowing fiercely at 100 knots for more than a day, and *Lady Alice* was struggling in rough seas with waves 16 to 23 feet high. Spray and mist were blowing off the waves so that it looked as if the ocean was smoking, and visibility was greatly reduced. At 8:00 A.M. Gallimore looked up and saw a huge wall of water bearing down on *Lady Alice*. From his view in the wheelhouse he could not see the top of the wave. There was no time to do anything. The wave crashed down on top of the wheelhouse, driving the vessel underwater. The force of the wave ripped off the radio and radar antennas on top of the wheelhouse, bent down the steel plate "eyebrow" that ran around the top of the wheelhouse above the windows, blew out all seven of the heavy lexan windows in the front of the wheelhouse, and smashed out the wooden wall that made up the back side of the wheelhouse. After several agonizing seconds the boat popped back up to the surface.

When the wave hit, Gallimore was hurled backward into the chart table, badly bruising his thigh and back as water poured into the wheelhouse through the window openings. He was holding on to the wheel at the time the wave hit. The force was so great that he literally ripped the two spokes he was holding right out of the wheel. The crewman in the wheelhouse with him was thrown down with such force that he suffered two fractured vertebrae. Saltwater shorted out all of the instruments. Gallimore threw the main breakers, which prevented more damage. All of the instruments were gone except the fathometer—including radios. Fortunately, the engines continued operating.

Gallimore realized that *Lady Alice* had been hit by a rogue wave. They were in the Labrador Current at the time and the location must have contributed to the height of the wave that hit them. Years later, after he had moved his fishing fleet to Hawaii, we discussed the incident during breakfast on Pier 38 in the Honolulu Harbor. "We were lucky there was just one wave," said Gallimore. "If another one had come, we probably wouldn't have made it."

Twenty-five years later, *Lady Alice* is still catching fish. At Gallimore's invitation, I went aboard as she was docked in Honolulu between trips. The crew was busy rigging lines and provisioning the boat for the next trip. Gallimore introduced me to Captain Chase Blinsinger, who is currently running the boat. We went up to the wheelhouse, where Gallimore pointed out the damage. I could see the joint where a 6-foot-long section of steel plate had been welded into the eyebrow at the roofline of the wheelhouse. All of the wheelhouse windows are now reinforced with stainless steel frames and inside, behind the lexan, horizontal stainless steel bars ensure that the windows cannot be blown in again under the force of a wave.

How high was the wave? Gallimore does not know—he could not see the top from where he stood in the wheelhouse. But with the significant wave height running around 20 feet at the time, theory says that a rogue wave would have been at least 44 to 48 feet high. To top the radar antennas with enough force to rip them from the steel mast where they are bolted—not just wash over them—the wave had to be 40 feet or higher. To drop down on the top of the wheelhouse with considerable force—enough force to momentarily push the vessel underwater and bend down the edge of the steel deck over the wheelhouse—would have taken tons of water striking with enormous impact. It would take a wave of that height to explain the damage that occurred.

Blinsinger showed me the satellite phone on board and the computer he uses to get the latest weather forecasts and current direction and velocity as well as sea surface temperatures—vital information for locating schools of tuna. The information is provided by a wireless interface to a commercial service and updates are available every two hours.[14] With the click of a mouse, this service provides complete weather data, satellite images of every ocean, chart overlays on satellite images, ocean water temperature, significant wave heights and swell direction, ice locations, storm warnings, and more. All of the boats are also equipped with global positioning satellite systems so that their positions are known accurately at any time. Through another commercial service, this information is uploaded to a satellite and downloaded to the Internet. Gallimore can sit in his home overlooking Kaneohe Bay on Oahu, turn on his computer, and with a click of his mouse know the positions of all six boats, current speed, and direction of travel. Combining this information with the latest weather forecasts gives him—and the boat crews—an extra margin of safety that never existed in the early days of offshore fishing.

I asked Gallimore if he found the weather in the Pacific better than the Atlantic. "Generally, yes," he said. "And we can fish all year round. But even here we get storms. *Lady Alice* was caught out in Hurricane Iniki, in the channel between Oahu and Kauai. She rode out the storm pretty well, although the wind was blowing so hard we got some leakage around those stainless steel frames we put on the wheelhouse windows."

I remembered the two times I made the crossing between Oahu and Kauai in the *Kuu Huapala,* with Captains George Paxton and Karl Adams, on our way to fish for marlin around Kauai and Nihau. We ran all night to get across in seas 6 to 8 feet high—no storms, just the normal sea state for the Kauai Channel. To the north and south of the islands the depth is more than 2,000 fathoms, while in the channel the bottom rises up and is only 400 or 500 fathoms deep and the islands act to funnel big swells from the northeast into the channel. It seems it is always rough out there. At one point we had a big ice chest with a couple of hundred pounds of ice. It got loose as the boat rolled in the seas and gave new meaning to the term "loose cannon," until we were able to get it lashed down again. Personally, I wouldn't want to be in the Kauai Channel in a storm, much less in a hurricane! I thanked Gallimore and Blinsinger for their time, and wished them good fishing and good weather.

"Don't run into any rogue waves," I told Gallimore. "Yeah, right," he said. "Once in a lifetime is enough."

Not long after my visit, a damaged ship limped into Honolulu. It was the 590-foot-long cruise ship M/V *Explorer,* with 700 college undergraduates on board—really a floating college, complete with a library, nine classrooms, and a computer lab. The vessel departed Vancouver on January 18, 2005, en route to South Korea. In the early morning hours of Thursday, January 27, it was about 565 nautical miles south of Adak, Alaska, when it was hit by a rogue wave 50 to 70 feet high, according to various news reports.[15] The wave rocked the vessel, tossing students from their beds, causing computers to fall, and dumping books from the library shelves. It hit the bridge and pilothouse, broke windows, flooded the bridge, and injured two crew and several passengers. As in the case of the *Bremen,* saltwater shorted out the instrumentation and controls and left the vessel without power for a period of time. To determine how high the wave really was, I contacted officials at the Semester at Sea Program and was told that the top of the bridge was 85 feet above the waterline, the floor of the bridge about 75 feet above the waterline. Since the vessel was rolling at the time of the incident, with a 30 to 40 degree heel angle, it would have taken a wave 57 to 65 feet high to reach the level of the bridge.

Jen Hanson studied at Hartwick College in Oneonta, New York, majoring in psychology. One day while searching the Internet for study abroad programs, she stumbled across the Semester at Sea Program and decided to participate. What follows is her recounting—as told to me—of the days immediately preceding and just after M/V *Explorer* was hit by the rogue wave.

"Sea conditions got rough about two days after we left Vancouver. A lot of people were seasick, although I was not. One of the faculty members had to stop in the middle of a lecture and run outside because of seasickness. The boat was rolling strongly, so you had to be careful how you walked, always holding on to something to keep from banging into the walls. We were not allowed on deck until the sixth and seventh day out. At this point the crew said we were experiencing 15- to 30-foot-high waves. By January 25, it was a little rougher. Chairs would roll back and forth; beds and furniture would slide in our cabin. I could see wave tops at the level of my window on the third deck. We were on the port side, near the stern of the M/V *Explorer*.

"On the morning of January 26, my roommate and I awoke around 1:00 to 2:00 A.M. because the boat was rolling so much that our beds were sliding around. It didn't seem safe in the room, so we went out in the hallway to sleep in our sleeping bags where we could kind of brace ourselves. Unable to sleep, we went up to the sixth deck, where the computer lab was located; there we saw lots of broken glass, computers crashing on the floor, chairs tumbling over. In the dining room, silverware was dumping out on the floor, furniture flying around. In the gift shop the glass doors were shattered, but the glass was still in the frame. Then we realized that sea conditions were definitely rougher than anything we'd seen up to that point. We went back to the third deck hall but after a while gave up and decided to try our cabin once more and see if we could get some sleep. It was around 4:00 to 4:30 A.M. We opened the door, were in the room a few minutes, then the boat took a huge roll to the port side (throwing us in the direction of the window). At this point we left the room and returned to the hallway.

"The biggest wave hit the bridge, on the seventh deck. The captain later said that the vessel was pitching up and down when the wave hit the bridge. He said it was 45 feet high. He told us that the ship was not in a storm but was experiencing waves from two storms, one on either side of the vessel, and that the swell from these distant storms came together and created the rough conditions. On the bridge, the wave knocked out one window and damaged the electronics. This caused all four engines to stop operating. Someone came on the P.A. system and told us to put on life jackets and move into the halls.

"After about 20 to 30 minutes the crew was able to get one engine started, and a while later, a second engine. The ship wallowed in swells with no steerage for a time. We had a hard time opening cabin doors with the ship heeled over. Two crew members were injured, one with a concussion—not sure about the other. One faculty member had a broken hip. Initially there was some talk the ship would divert to Midway Island, but then it was determined that we could make Honolulu. Seas continued to be rough for two more days, and then the last four days into Honolulu were better."

Even the largest vessels are not immune to rogue waves that arise during storms. On April 12, 1966, the *Michelangelo* had a frightening encounter with a rogue wave. This 902-foot-long passenger vessel was about 700 nautical miles east of New York in the midst of a storm. The significant wave heights were 20 to 30 feet. Suddenly she was hit by an extreme wave of such violence and force that the steel bow of the ship was damaged, steel railings on the upper decks were torn free, the bulkheads under the bridge were crumpled, and windows in the bridge were broken. Large quantities of seawater poured into the ship through damaged bulkheads. The bridge windows, of 1-inch-thick glass, were located 81 feet above the waterline. Three people were killed and 13 others were injured.[16]

In 2002, I had the pleasure of sailing in a chartered sailboat from St. Martin Island to St. Barthélémy, and then to Anguilla. The weather was perfect, the sailing wonderful, and we enjoyed diving and some great fishing as well. I remember one evening in particular, when we were anchored in Cove Columbier on St. Barts—you could not possibly find a lovelier spot. I was awake in my bunk, a hatch open overhead, the stars bright above me. In the morning I awoke early to do some writing, but finally had to close the hatch when a short refreshing squall passed by. Glancing around the islands, I saw that they were verdant, flourishing. The scars of Hurricane Luis had disappeared.

August is hurricane season in the Atlantic, and the Cape Verde Islands off the west coast of Africa are known to be a spawning ground for hurricanes. Near the end of August 1995, a circulation of low clouds was first noticed in the Cape Verde area. Soon it began moving westward and Hurricane Luis was born. The track next swung to the northwest and passed over the Leeward Islands. Antigua, St. Barthélémy, St. Martin, and Anguilla were affected by sustained winds as high as 115 knots in the eye wall of the hurricane. At least 17 people died and damage was estimated at $2.5 billion.

Luis was a very large hurricane, the inner eye diameter equal to 40 nautical miles. As it recurved north, it impacted Nevis, St. Kitts, and the northernmost British Virgin Islands, finally passing about 200 nautical miles west of Bermuda on its way to Newfoundland. On September 6 and 7, the hurricane passed over the ship *Teal Arrow*, which reported sustained winds of 99 knots with gusts of 125 knots. Wave heights were estimated at 50 feet.[17]

Queen Elizabeth 2 is a 963-foot-long passenger ship launched in 1969. On the morning of September 11, 1995, she was about 200 nautical miles south of eastern Newfoundland and 120 nautical miles from the center of Luis. It was at this point that the *Queen Elizabeth 2* encountered a 95-foot-high extreme wave, prompting Captain Peter Warwick's comment that "it was a giant wall of water, like running into the white Cliffs of Dover." A Canadian data buoy not far from the ship's location reported a peak wave height of 98 feet at about the same time.

Near the end of World War II a series of typhoons did more damage to the U.S. fleet than the Japanese Navy could inflict at that time. On December 18, 1944, a typhoon struck Admiral Halsey's Task Force 38 while engaged in naval operations around 300 nautical miles east of Luzon. The storm was accompanied by 100-knot winds and seas 70 to 80 feet high. Three destroyers capsized and sank (the *Hull, Monogan,* and *Spence*), although sister ship *Dewey* survived, despite rolling at times as much as 70 degrees. Nine other vessels were seriously damaged, three carriers sustained fires, and 146 aircraft were lost. There were 790 fatalities; 80 persons were injured.

Several more typhoons harassed the fleet as it neared Japan. One of the more unusual incidents involved the heavy cruiser *Pittsburgh (CA-72)*, launched in 1945. She was 675 feet long. On June 4, 1945, the *Pittsburgh* found herself in the middle of a growing typhoon. By June 5, winds had reached 70 knots and waves were 100 feet high. Wind tore her starboard scout plane from its catapult and crashed it onto the deck. Next she was struck by a huge wave, causing her second deck to buckle, and 90 feet of the bow structure was ripped upward and torn free from the vessel. In the same storm, waves tore away supporting structure under the forward flight decks of the carriers *Bennington* and *Hornet*, causing the forward section to collapse. The carriers continued flight operations by reversing the launch direction and flying planes off the stern rather than the bow.

The challenge now for the *Pittsburgh* was to maneuver to stay afloat and avoid being rammed by the drifting bow section. The crew kept the vessel quarter-on to the mountainous seas while emergency repairs were made and the forward bulkhead was reinforced. She was finally able to creep back to Guam for repairs. Meanwhile, the derelict bow section, now nicknamed the SS *McKeesport* (after a suburb of Pittsburgh), was eventually salvaged by another vessel and brought to Guam. The *Pittsburgh* was repaired and later served in the Korean War.
18

Rear Admiral Joe Barth served on the USS *Hancock* during the early 1970s as an air wing commander of five squadrons of naval jet aircraft (three squadrons of A-4s, two of F-8s). During his career as a naval aviator, Admiral Barth made a thousand carrier landings. He subsequently served as the captain of the *Milwaukee* (AOR2), a 650-foot-long supply ship with a crew of about 250 men, and then in the mid-1970s became the captain of the aircraft carrier *Forrestal*. While bringing the *Milwaukee* back from the Mediterranean in October 1975, navy weather forecasters advised of a coming storm and rerouted the ship east of Bermuda—and into the storm. The storm got progressively worse, with storm force winds of 48 to 63 knots and 40-foot-high seas. One wave came clear over the bow of the ship and hit the bridge, about 75 feet above the waterline. The bridge was not damaged, but equipment on the bow of the vessel was ripped off and lost overboard.

Continuing, Admiral Barth told me that he recalled an incident on the aircraft carrier *Valley Forge* (CV45) in a storm around 1957-1958. This was during the time before carriers were equipped with a "hurricane bow"—the steel plate flaring that rises up from the bow to the flight deck. Before, the flight deck extended forward and there was an open space beneath it. In this particular storm, the *Valley Forge* was pounded and one wave ripped off the front port-side section of the flight deck with the catapult. The damaged area was hanging over the side and beating on the ship, threatening to do further damage, and had to be cut away. The ship made it into port and was successfully repaired.

"I guess you can say I've been lucky," he said. "You really have to respect the tremendous energy present in large waves."

If anyone knows about rogue waves, it is the masters of oil tankers that cross the major oceans. In this regard I was fortunate to meet Captain Mike Miller and Captain Jan Jannsen, both of whom drive tankers carrying 1 million to 3 million barrels of oil.

Miller told me that in 23 years at sea—8 as a shipmaster—he'd experienced storms and heavy seas, but only one rogue wave. It was on a trip from Valdez, Alaska, to San Pedro, California—a trip that normally takes about a week, depending on the weather. On this route, the spots where you would normally expect rough seas are rounding Point Conception and passing Cape Mendocino.

Miller: "It was in 1982 or 1983. After leaving Valdez, we were heading south through the aftermath of a winter storm, with two more on the way, a veritable string of pearls. Although the engine was turning RPMs equivalent to 9 knots, the *Atigun Pass* was only making about 4 knots.[19] Suddenly the vessel dropped into a deeper than usual trough and a huge wave broke over the main deck of the 905-foot-long tanker. The impact with this wall of water literally brought the tanker to a dead stop. Any of the 28 crew members who happened to be standing or walking around were thrown forward. In its fully loaded condition, the tanker deck was 19 feet above the waterline, and the bridge was an additional 50 feet above the deck. The wave flowed over the deck and rose up to the level of the bridge, before finally washing back into the sea. The tanker shuddered, rose, and continued on its way without further incident. Fortunately, we had no major damage. However, if you'd told me there were 69-foot-high waves out there, I wouldn't have believed you—until I experienced one myself."

Jannsen stated that he'd seen "sky-high waves, total green water," but none that he felt was truly a rogue wave. The big waves that he'd seen in his experience came from storms. Once, when leaving Ireland in a 197-foot-long anchor-handling tugboat, he encountered gale-force winds. He described driving into "a hollow sea, green water over the bridge, the fish looking in at me. . . ."

Jannsen is also familiar with South Africa and the Agulhas Current. He'd never experienced a rogue wave personally, but told me about seeing masts down, rails twisted and torn away, lifeboats ripped from their supports and washed overboard—all the result of large waves.

Although the focus of this book is on oceans, I would be remiss not to say something about large waves in our inland freshwater seas—the Great Lakes. Commencing in November, the Great Lakes are susceptible to terrible storms. Hundreds of ships have been lost on the lakes since they became a favored path for the transport of freight and passengers. The largest lake, Lake Superior, is about 350 miles long and 160 miles wide, with an average depth of 483 feet and a maximum depth of 1,332 feet. Lake Michigan, the second largest, is about 300 miles long, 118 miles wide, and about half as deep as Lake Superior.

Given the size of the lakes, the waves that can be created by winter storms are fetch limited. Table 2 shows that a steady wind of 30 knots blowing for 23 hours over a minimum fetch of 243 miles (about the length of Lake Michigan) will produce waves that are 19 to 34 feet high. These waves would theoretically have a period of 12.5 seconds, a speed of 38.2 knots, and a wavelength of 800 feet. Since the waves are fetch limited, you would not expect to see 100-foot waves in the Great Lakes. However, there are two complications: First, storms can come up quickly; second, in the deepest parts of the lakes, the depth is deep relative to the wavelength. As the waves move into shallower water, they slow down, and the wave heights grow as the water "piles up." Also, with strong winds blowing, breaking waves can be expected. Historically, wind speeds in excess of 87 knots have been recorded, so in major storms you would expect to encounter "fully developed seas."

This is what happened on November 18, 1958, when the *Carl D. Bradley*, a 638-foot-long ore carrier left a dock near the southern end of Lake Michigan and proceeded north, heading for Lake Huron. The wind blew heavily that night and by the next day had reached 52 knots, creating large waves that broke over the bow of the ship—one of the two surviving crewmen later estimated the wave heights as 30 feet, which tends to agree with fully developed sea conditions that might have occurred. Just before darkness arrived, the crew heard several loud noises and realized the vessel was breaking up. They had time for a quick Mayday before the *Bradley* broke into two parts. Within minutes, the two halves of the ship sank and the crew of 35 was in the freezing water. Four men managed to make it to a raft that had floated free before the *Bradley* sank. Two of them died during the long night, but two managed to survive and were rescued by the coast guard in the morning.[20]

The largest freighter ever lost on the Great Lakes was the *Edmund Fitzgerald*. Another victim of a sudden November storm, the *Fitzgerald* sank on November 19, 1975, with a crew of 29 men on board. There were no survivors. The tragedy was memorialized in a well-known song by Gordon Lightfoot, "The Wreck of the Edmund Fitzgerald."

What happened was this: After taking on a load of 26,000 tons of taconite (iron ore) at Superior, Wisconsin, the *Fitzgerald* headed northeast and then east along the length of Lake Superior toward the Soo Locks and Sault Ste. Marie, bound for Detroit. About 15 miles behind the *Fitzgerald* was the *Arthur M. Anderson*, another ore carrier that maintained radio and radar contact with the *Fitzgerald* until she sank.

The storm came over the lake near Marquette and blew northwest toward Canada. The Soo Locks reported peak winds at 87 knots and the locks were shut down. Captain Jessie B. Cooper of the *Anderson* reported that waves 8 to 12 feet high were washing over his deck. He later told Fred Shannon (a researcher investigating the sinking of the *Fitzgerald*) that one wave hit the stern quarter of the *Anderson* and damaged a lifeboat that was supported 27 feet above the waterline. In its loaded state, the cargo deck of the *Fitzgerald* was only 12 feet above the waterline.

Around 7:00 P.M. the *Fitzgerald* disappeared from the *Anderson's* radar screen. The last word from Captain McSorely on the *Fitzgerald* was: "We're holding our own." After the vessel sank, underwater inspection revealed that she had broken in two at about the one-third, two-thirds point. Various theories have been put forth as to the cause of the breakup, including the possibility that some of the *Fitzgerald's* hatches were not sealed, allowing the vessel to take on water.

However, Captain Cooper made an interesting point in his recounting of the storm. He stated that around 7:00 P.M. the *Anderson* took two huge waves in rapid succession, the first flooding the deck, the second throwing green water up on the bridge deck. He speculated that these waves (remember, they were probably traveling at around 38 knots) overtook the *Fitzgerald* a few minutes later and, if joined by a third rogue wave, might have overpowered the struggling vessel—evidence of an occurrence of the Three Sisters in a Great Lakes storm.[21]

CONTINENTAL SHELVES AND SHALLOW SEAS

The third source of extreme waves is wind-driven waves piling up in shallower waters such as those of the continental shelves or the North Sea, where depths are 100 to 650 feet. This is similar to the phenomenon of waves on a beach—waves increasing in height as the water becomes shallow. An extreme wave can cause flooding in coastal areas when a wall of water inundates the shore for several minutes, or even tens of minutes, or occurs as a low-frequency wave train with a period of 30 seconds. Extreme waves pose a hazard to marine structures other than ships, examples being ports, harbors, piers, and offshore oil platforms. Notable examples include the loss of Texas Tower 4 in the North Atlantic and the loss of oil platforms in Canada and the North Sea.

In the 1950s, the U.S. Air Force planned to build five offshore platforms along the East Coast of the United States. Because of their similarity to offshore oil platforms in the Gulf of Mexico, they were called "Texas Towers." The towers were built as part of the U.S. early-warning air defense system to detect the approach of enemy aircraft or missiles. They contained radar and communication equipment and a total crew of around 50 men. This position was expected to give an additional 30 minutes of warning in the event of enemy attack. By 1960, three towers were operating; the plans to build numbers 1 and 5 were eventually shelved.

Of the three operating towers, U.S. Air Force Texas Tower 4 was built in the deepest water. It was located at an offshore location called Unnamed Shoal, 73 nautical miles east of New York City, in water 180 feet deep. The operating decks were 60 feet above the waterline. The design conditions were for winds of 109 knots and for breaking waves 35 feet high.

On September 12, 1960, the tower was hit by heavy seas during Hurricane Donna and was evacuated. Winds of 115 knots and breaking waves exceeding 50 feet occurred. When the storm subsided, the crew returned and found that the tower had been damaged. Plans were made to make major repairs in the spring. A skeleton crew of 28 men was left on the tower to start some repairs; full evacuation was scheduled for February 1, 1961. Unfortunately, before this evacuation took place another storm occurred on January 14-15, 1961, creating conditions too rough for evacuation. The tower was rocked by 74-knot winds and wracked by a succession of large waves 35 feet high, causing it to collapse with the loss of the entire crew.[22]

The 1980s were particularly cruel to the offshore oil platform industry. Sometimes it is not the "big one" that causes the damage, but rather the cumulative effect of many waves impacting a structure. This is what happened to the *Alexander Kielland*, a floating drilling platform owned by Phillips Petroleum. The platform, named for a famous Norwegian writer, was used for several years as a drilling platform at the Ekofisk field in the North Sea. Later it was converted into a floating hotel for oil field workers. On March 27, 1980, a fatigue failure caused the loss of the support bracing connecting one leg to the platform. The platform was held in place by anchors, and the multiple legs were used as flotation devices. When the leg failed, the platform capsized, and 132 of the 212 persons aboard drowned. Earlier, I described my own experiences in the North Sea at Ekofisk, including time spent on platform Two-Four-Delta, which was struck by an extreme wave while under construction. This was several years after the *Alexander Kielland* capsized, but the disaster was still a poignant memory for the people with whom I worked.

Since 1980, extensive monitoring of the sea state has been made at Ekofisk and the occurrence of extreme waves 66 feet or higher documented on a number of occasions. Damage has occurred, sometimes resulting in the loss of life. The adjoining oil field at Draupner collected a historic "first," during a storm on January 1, 1995, when an extreme wave of 84 feet was actually recorded, as depicted in Figure 18.[23] An analysis of the sea state at the time this wave occurred gave a significant wave height of 39 feet, so the extreme wave criterion (the ratio of these two numbers) was 2.15 or close enough to the value of 2.2 to characterize the wave as an extreme wave. The crest height was 60 feet with the trough depth only 24 feet, demonstrating the asymmetry of an extreme wave. The steepness of the wave was 0.06, about twice that of the significant waves.

CONSTRUCTIVE INTERFERENCE (SUPERPOSITION)

Another cause of extreme waves is the random superposition of two or more wave trains in a confused sea—wave trains that happen to have one or more wave crests that occur at the same instant, combining to create an extreme wave. Superposition can occur in "normal" seas when there is no apparent storm or unusual winds. This was the situation described earlier in this chapter, as we were sailing to Catalina Island, although the resulting wave could hardly be called extreme.

Ernie Barker is a former merchant mariner who dissolved a successful law practice when in his 50s, bought a 41-foot Kettenburg yawl called *Nepenthe*, and set sail for Australia. My good friend and neighbor Keith Garrison now owns *Nepenthe*, and he put me in touch with Barker.

As a kid, Barker always wanted to go to Australia. After serving nine years in the merchant marine, he still hadn't made it to Australia. In 1950, upon departing from Yokohama, Japan, while still in the merchant marine, they ran into a gale that came crashing in on an entire

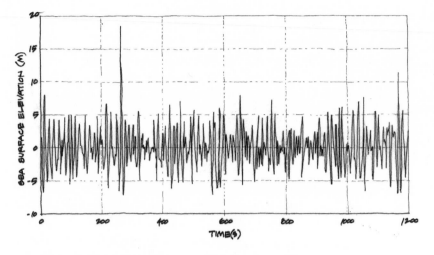

FIGURE 18 Draupner oil platform wave height, January 1, 1995.

fleet of small Japanese fishing boats. The radio waves were suddenly filled with Mayday calls. Barker and his crewmates rescued the crew of one boat, losing a lifeboat to a rogue wave in the process. That was the first time he'd heard the term used. It stuck with him.

He bought *Nepenthe* in 1979 and began outfitting her for cruising, sailing days and weekends in Southern California whenever he had a chance. In Chapter 2, I mentioned sailing to Santa Barbara Island in my first boat; subsequently, I've been there in *Dreams* perhaps a dozen times.

Barker told me that he'd been to Santa Barbara Island also; what impressed me was that his trip to this island was the longest ocean passage he'd made on a sailboat before setting off for Australia! He left California on October 14, 1982, and made his first landfall as planned at Pago Pago, American Samoa, 54 days later. It was a little longer than going to Santa Barbara Island.

After successfully reaching Australia, Barker was single-handing *Nepenthe* between Bundaberg, Australia, and Noumea, New Caledonia, somewhere around the Tropic of Capricorn and longitude 160 degrees east, in the Tasman Sea. In his own words, here's his story.

"I was beating on a starboard tack with a storm jib and fully reefed mainsail. Winds were 40 knots, gusting to 60; but all seemed well. I was sitting in the cockpit, dead tired, approaching the end of a second day without sleep, when a wall of water dropped into the cockpit. It flattened out around my ankles on the cockpit floor, about 8 inches deep. That brought me to full alert. I scrambled to slide a second board in the companionway doorway. Before I could get the next board in place, a second mountain of water washed in from the port stern quarter. This time it washed over the companionway board and down into the cabin. I quickly got the second and a third board in place and then grabbed a five-gallon bucket and started bailing out the knee-deep cockpit. Once I had it under control, I put one leg over the companionway boards and reached below for a towel to stuff into the door boards to make them a little more water-tight. At that precise moment, the third wave hit starboard with such force that I was thrown in a heap onto the cabin floor. I staggered to my feet, my left arm numb and useless for several days, but fortunately not broken.

"This was my introduction to rogue waves. It wasn't an introductory course, with a single wave; I had a graduate course, with the full Three Sisters' treatment."[24]

Barker survived the waves this time, and several more times as well, and brought *Nepenthe* safely back to California. He wrote a journal of his trip, fittingly titled *Rogue Wave*, from which this excerpt is taken. At several points in his journal, Barker sums it up by saying, "The devil lives in the Tasman Sea, and he'll beat you to death with his rogue waves if you don't watch out for him...."

You can meet Barker if you care to travel to San Pedro (the Port of Los Angeles) and board the SS *Lane Victory*. The *Lane Victory*, now a national monument, is where Barker works as a volunteer steam engineer. The *Lane Victory* is a fully restored, 455-foot-long World War II victory ship that sails from San Pedro on Saturdays and Sundays during the summer. Enjoy a cruise to Catalina, along with a simulated air attack by a squadron of World War II planes that come screaming out of the sky! Not only can you meet Barker, you can support this historic merchant marine steamship—a worthy cause.

In the 20-year period from 1980 to 2000, more than 200 super-carriers—bulk carriers, combination carriers, and crude oil tankers more than 656 feet (200 meters) long—were lost to storms.[25] In a number of cases, the cause was a giant wave that hit the vessel suddenly without warning. Sometimes these waves appeared in the midst of a storm; on other occasions they occurred when seas were moderate. It is not just bulk carriers and tankers that have been hit. The preceding paragraphs include only a small sample of known incidents. Virtually all types of vessels—oil tankers, bulk carriers, container ships, passenger liners and cruise ships, fishing boats, even aircraft carriers and naval ships designed to withstand bomb blasts—have fallen victim to extreme waves at some time or another. (As part of my research I compiled a listing of more than 70 such incidents.)

HOW HIGH CAN THEY GET?

In addition to the wave heights seen by shipboard observers or personnel on offshore oil platforms, wave heights have been measured and recorded for a period of years at various locations by instruments. By examining large numbers of records that cover a long period of time, it is possible to determine, on average, how frequently a wave of a certain height can be expected to occur. In technical jargon, this is referred to as the *return period* (measured as 2, 5, 10, 50, or 100 years), meaning that on average a wave of that particular size can be expected once every 2, 5, 10, or so many years.

Wave measurements have been made by weather ships, nine of which are located in the North Atlantic along the routes most frequently traveled by merchant shipping between Europe and North America. In the Atlantic, the highest significant wave height expected to occur once in 2 years, ranged from 23 to 39 feet; once in 50 years, it could be as high as 43 to 75 feet. In the Pacific, the range for once in 2 years was 20 to 36 feet, and for once in 50 years, 36 to 66 feet.

These are the values for the significant wave height. However, remember that the significant wave height is the average of the top one-third of the highest waves. It is *not* the extreme wave. In the Atlantic, a probability analysis would show that the extreme wave could be as much as 131 feet and in the Pacific as much as 118 feet.[26] In the North Sea, predictions of maximum wave height give 111 feet. The highest measured to date is the famous 84-foot "New Year's" wave of January 1, 1995, recorded at the Draupner oil platform.

At one time, measurements of wave heights in the North Atlantic showed that average wave heights seemed to be increasing, a pattern also observed in the last 15 years. Today this is a matter of scientific debate. Is this a trend, or does it just reflect the natural long-term variation of weather patterns? Weather seems to be moving north, bringing more intense storms to the north. This could be due to migration of warmer water north (known to advance and retreat periodically), to expansion or shrinkage of the ice boundary, or to storm patterns that "huddle" ice packs together.

Surprisingly, 100-foot-high waves have recently been measured off the west coast of Canada. Commencing just south of Vancouver Island and continuing north to the British Columbia–Alaska border, the Canadian government had 16 weather buoys. Some were placed between Vancouver Island and the mainland, others in Hecate Strait, and a number offshore on the west side of Vancouver and Queen Charlotte islands. During a storm on December 10, 1993, the East Dellwood buoy, located 61 nautical miles west of Cape Scott, recorded an extreme wave of 101 feet. The wave occurred at a time when the buoy recorded significant wave heights of 40 feet. Earlier, on December 20, 1991, the south Hecate buoy measured a wave that was 100 feet high. On December 13, 1992, the West Dixon entrance buoy was swamped by an 85-foot-high wave. It seems as though the waters around Vancouver and Queen Charlotte islands would be a good place to avoid during the month of December.[27]

One of the earliest reliable measurements of a giant wave was made on February 7, 1933, by the U.S. Navy oil tanker *Ramapo* while steaming in the North Pacific. The USS *Ramapo* (AO-12) was a fleet oil tanker capable of carrying 70,000 barrels of fuel oil to refuel navy ships at sea. The vessel was launched in 1919 and initially saw service in the Caribbean, carrying fuel from Texas to Guantanamo Bay, Cuba. In the 1920s, she was transferred to the Pacific, where she serviced ships of the Asiatic Fleet, carrying oil from San Pedro, California, to the Philippines and China. *Ramapo* was in Pearl Harbor when the Japanese attacked, returned fire, may have shot down one plane, and miraculously sustained no damage. Later in the war, during a bad storm near the Aleutian Islands in December 1942, the *Ramapo* was credited with rescuing the entire crew of the USS *Wasmuth,* a minesweeper that sank when depth charges broke loose on deck during heavy weather and exploded. However, the *Ramapo* is best known for providing the first carefully documented record of an extreme wave.[28]

In February 1933, while on its usual run from Manila back to California, *Ramapo* was caught up in a ferocious seven-day storm. Winds blew at speeds of 60 knots and higher. An officer on watch on the bridge observed a wave approaching the stern of the vessel at a height just above the crow's nest on the main mast, while the stern of the vessel was sunk in the trough of the wave. When the geometry of this sighting was worked out using the known height of the mast and the length of the vessel, the wave was found to be 112 feet high (see Figure 19). The period of the wave was measured as 14.8 seconds and its wavelength calculated to be 1,100 feet. Since the *Ramapo* was only 480 feet long, she was able to ride through the wave without having her back broken while suspended on a crest or having her stern buried under tons of water.[29]

THE MAXWAVE PROJECT

The Commission of the European Communities recently initiated a multination research project on extreme waves.[30] The MaxWave project goals are to develop a better understanding of extreme waves, determine their frequency of occurrence and the places where they are most likely to occur, and develop methods of predicting their occurrence. The project addresses extreme wave occurrence not only in oceans and the effects on shipping, but also in shallow and coastal waters and the effects on offshore oil platforms and ports and harbors.

MaxWave is divided into 10 tasks, commencing with the establishment of a database by investigating known extreme wave incidents and developing extreme wave statistics (wave heights, periods, frequency of occurrence) from buoy and satellite radar data from which it will analyze the regional distribution of extreme waves. Known ship accidents involving extreme waves also will be analyzed and, where pos-

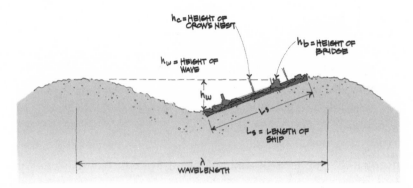

FIGURE 19 USS *Ramapo* wave height measurement scheme.

sible, compared to hindcasts of existing sea conditions at the time of the incident. By using the results obtained from the research, wave conditions and criteria for designing safer ships and offshore structures will be developed. The research is ongoing, but one early finding is that waves 66 to 98 feet high occur much more frequently than previously thought.

As part of the MaxWave team, Dr. Susanne Lehner and her colleagues used satellites and radar imaging to scan large areas of the oceans, looking for rogue waves. In 2001, they discovered 10 waves exceeding 80 feet in height. Due to limitations on the data sets and sampling frequency, this number probably underestimates the actual number occurring. Fortunately, many of these waves are short-lived—they are born suddenly, rise up, and collapse quickly without propagating for any significant distance. Unless a ship is in the vicinity, or a satellite overhead, they live and die in obscurity, somewhere in the vastness of the world's oceans.

Another extreme wave measurement is that described in Chapter 4, when Hurricane Ivan passed over an array of underwater sensors that detected a 91-foot-high wave.

BOTTOM LINE

In conclusion, there is ample evidence for the occurrence of extreme waves. As the examples listed above demonstrate, big waves can no longer be explained away as simply a sailor's exaggeration. (See Table 5).

After hearing or reading dozens of accounts of extreme wave incidents, I was impressed with the consistency of the descriptions. Inevitably, they all contain several of these phrases: "It came out of nowhere; it was a wall of water; it towered over the boat; it was as if we'd fallen into a hole in the sea."

Frightening images—but what do they portend for the future? Is it safe to say that modern ships are sufficiently robust to survive the worst weather that oceans can throw at them? Would you or I want to be on a vessel facing the Draupner wave?

TABLE 5 The Evidence for Extreme Waves

Description and Location (year)	Wave Height in Feet		Notes[a]
	Significant	Extreme	
Lituya Bay, Alaska (1958)	320	1,700	Landslide
Sydney-Hobart Race (1998)	40-60	140	Measured
Krakatoa, Indonesia	—	133	Volcano
Weather ship data (ca. 1980)			
Atlantic	43-75	131	Calculated
Pacific	36-66	118	Calculated
Offshore platforms			
North Sea	—	111	Calculated
USS *Ramapo*, N. Pacific (1933)	—	112	Measured
East Dellwood, N. Pacific (1993)	40	101	Measured
Ocean Ranger, N. Atlantic (1982)	—	100	Estimated
SS *Bremen*, S. Atlantic (2001)	—	98	Estimated
Submarine *Grouper*, Atlantic	Calm seas	98	Measured
Caledonian Star, S. Atlantic (2001)	—	98	Estimated
Athene, Indian Ocean (1977)	—	98	Estimated
Queen Elizabeth 2, N. Atlantic	—	95	Estimated
Hurricane Ivan, Atlantic (2004)	—	91	Measured
Queen Elizabeth, N. Atlantic (1943)	—	90	Estimated
Unnamed vessel, Atlantic (1960s)	—	85	Estimated
Draupner platform, N. Sea (1995)	39	84	Measured
Esso Nederland, Agulhas current	—	82	Estimated
MaxWave satellite study (2001)	—	80+	Measured
May 22 Chile earthquake (1960)	—	82	Runup
Ob, Southern Ocean (1955-1956)	—	82	Estimated
Michelangelo, N. Atlantic (1966)	20-30	81	Estimated
USS *Milwaukee*, N. Atlantic (1975)	—	75	Estimated
Tanker *Atigun Pass*, Pacific (1982)	—	70	Estimated
USS *Wateree*, Chile (1868)	—	69	Tsunami
Schiehallion, N. Atlantic (1998)	46	66+	Measured
Munchen, N. Atlantic (1978)	—	66+	Estimated
MV *Explorer*, N. Pacific (2004)	—	61	Estimated
Independence, N. Atlantic (1977)	25-30	55-60	Estimated
Teal Arrow, N. Atlantic (1995)	—	50	Hurricane
Texas Tower 4, N. Atlantic (1960)	—	50	Hurricane
Polar Star, N. Pacific (1985)	11.5	35	Storm
March 27 Alaska earthquake (1964)	—	33	Tsunami

[a]In compiling this table, in many cases where height is "estimated," the estimate is based on the height above the waterline where significant damage was observed. The likelihood is that this underestimates the wave height. N means north; S means south.

10

When the
Big Wave Comes:
Are Ships Safe Enough?

For various economic reasons, today's commercial vessels are becoming larger and larger. Meanwhile, increased shipyard competition has created intense pressure to reduce shipbuilding costs. To reduce the amount of steel required—and thereby reduce labor costs—designers are making use of more sophisticated analytical tools for ship design and are specifying higher strength steel, so that less steel is required and therefore less shipyard labor. The net result is a substantial reduction of structural safety margins. There is also less material to resist the weakening effects of corrosion, so maintenance and frequent inspection are even more important and may not always occur as frequently as needed. Several of the shipmasters I interviewed expressed this concern. They said, in so many words, "Ships are not being built as well as they used to be."

This fact has been known about tankers since 1991, when a select committee of ship design experts was convened under the auspices of the U.S. National Research Council to review tanker designs in the wake of the *Exxon Valdez* disaster. Among their findings was the following: *As newer design techniques were introduced, "safety factors" (design al-*

lowances for unknown factors) were reduced, in the desire to keep costs down and to get maximum deadweight for minimum draft."[1]

Now that we know that waves as high as 100 feet exist and occur more frequently than previously thought, what can be done to reduce the number of vessels breaking up or foundering and to reduce the number of crew lives lost every year? A wave of such height, with its steep towering face, could deliver an enormous impact—enough to break the back of the best-designed modern ship built in accordance with today's standards. Even though such waves occur infrequently, are vessels safe enough?

EVOLUTION OF MARITIME TRANSPORT

In 2005 the world's merchant fleet comprised 39,932 vessels of 300 gross tons or more with a total deadweight tonnage (dwt) of 909 million tons.[2] (See Table 6.)

New and larger vessels, especially tankers and container ships, are being built in record numbers. One of the determinants of vessel size is the depth and width of the Panama and Suez canals. The maximum beam allowed through the Panama Canal is 105.5 feet. This corresponds to a container ship carrying around 4,500 containers, or 4,500 TEUs (20-foot equivalent units), meaning a container nominally 20 feet long. Ships that can traverse the Panama Canal are referred to as "under Panamax" vessels. In the Suez Canal, vessels with a beam of up

TABLE 6 Distribution of Vessels in the World Merchant Fleet

Vessel Type	Number of Ships	Million Deadweight Tons	Percentage of Merchant Fleet Capacity
Tankers	10,126	368	40
Bulk carriers	6,347	319	36
General cargo	16,263	95	10
Container	3,220	99	11
Passenger	3,976	28	3
Totals	39,932	909	100

to 243 feet are allowed, but the allowable draft depends on the beam and ranges from 52.5 feet to 36 feet for a wider vessel. Currently, tankers as large as 200,000 deadweight tons can pass through the Suez Canal.

The size of crude oil carriers (tankers) evolved from 10,000 to 20,000 deadweight tons in the 1940s to 1950s to 100,000 deadweight tons in the 1960s and 1970s. At first, size was somewhat dictated by the dimensions of the Suez Canal and the depth of major ports. However, recurring crises in the Middle East spurred the development of larger vessels. In the 1990s and 2000 very large crude carriers (VLCCs) with deadweight tonnage of 200,000 to 319,999 tons and ultra large crude carriers (ULCCs) with deadweight tonnage of 320,000 tons and more entered the fleet. These tankers can transport crude oil from the Persian Gulf around the Cape of Good Hope to deep-water ports in Europe and the U.S. East Coast, or across the Indian and Pacific oceans to the West Coast of North and South America.

At the beginning of 2005 there were 10,126 vessels in the tanker fleet, 7,650 of which were engaged in transporting crude oil. The average age of these vessels is 17.5 years, but it is dropping as the new double-hulled vessels enter the fleet and the older ones are retired. In 2003, the tanker fleet transported 1.7 billion metric tons of crude oil— the equivalent of roughly 12 billion barrels of oil, or nearly two-thirds of the oil consumed each year.[3]

The largest tankers are too big for most U.S. ports, so oil is offloaded at sea through a process known as *lightering*. For example, the *TI Europe*, a 442,470-deadweight-ton ULCC, 1,246 feet long, picked up 3 million barrels of crude oil in the Persian Gulf, crossed the Arabian Sea, then traveled southeast through the Indian Ocean, past tsunami-ravaged Sri Lanka and the coast of Sumatra, along Java, through the Lombok Strait (between Bali and Lombok islands), northwest through Makassar Strait between Borneo and Celebes, past Mindanao, and into the Pacific Ocean. This route follows or crosses the Philippine Trench and the Mariana Trench—some of the deepest waters in the world's oceans. A ship lost at these depths will never be found. Beyond Guam, the course is past Hawaii and on to Southern California—a journey equivalent to more than halfway around the world.

Accompanying Captain Jan Jannsen—on board the *Cygnus Voyager*, a double-hulled tanker with a capacity for 1 million barrels of oil—I observed a typical lightering operation.[4] The *Cygnus Voyager* was launched in 1994, is 900 feet long, and has a 160-foot beam—too large for the Panama Canal. She carries a crew of 30. On the bridge we were about 90 feet above the waterline. Looking toward the bow, the deck of the vessel stretched out—as long as two football fields laid end to end (but wider)—and was covered with a maze of pipes and valves for filling and emptying the various oil tanks. There were also a number of winches for mooring, two cranes, and four large sets of davits for lowering huge rubber fenders to protect the vessel when it came alongside *TI Europe*.

Jannsen, as lightering master, assisted the masters of both vessels in bringing them together. Using a handheld radio to communicate with both helms, he stood on the port bridge wing of the *Cygnus Voyager* and guided the two tankers onto the same heading at the same speed. Then he patiently brought the two huge vessels closer and closer together until the crew was able to shoot the first mooring line across. Finally, 12 mooring lines connected the two vessels. The mooring process took four hours. All night long the two vessels crept along at slow speed, while 1 million barrels of crude oil were transferred. At 3:00 P.M. the next day, Captain Jannsen gave the order to drop the mooring lines, and the *Cygnus Voyager* separated from *TI Europe* and headed north for the refinery. This process had to be repeated two more times before *TI Europe* could finally turn west and begin the 30-day trip back to the Gulf.

Bulk carriers carry coal, iron ore, grain, bauxite (aluminum ore), and other bulk products. Current vessels have capacities of 200,000 deadweight tons or more, and lengths of 900 to 1,100 feet. They have long, flat decks, hatches opening into the holds or storage spaces belowdecks, and an elevated bridge near the stern of the ship.

Bulk carriers and combination carriers (OBOs) have had a high loss rate since first introduced in the 1970s. In round numbers, from 1970 to 2000 at least 100 bulk and combination carriers have been lost, and more than 1,000 crew members have been killed. Some sinkings are due to human error or collisions, but a number seem to be due to preventable causes. It is estimated that 10 to 20 bulk carriers are lost annually due to structural failures.[5] This is clearly an unacceptable rate of loss and, presumably, something that can be remedied, as evidenced by a postmortem of the *Derbyshire* disaster discussed later in this chapter.

General cargo ships are the third most numerous type of vessels in the global fleet, but their numbers are declining and more are being sent to the ship breakers (scrap yards) each year than are being built each year. The decline in their numbers is due to the fact that greater volumes of goods are being carried by bulk carriers and container ships.

Container ships are among the largest vessels sailing the seas today, second only to the largest tankers. They also constitute the fastest-growing segment of marine transport. From 2000 to 2005, nearly 900 new vessels were added to the fleet. More than 300 vessels have a capacity of 5,000 TEUs and 15 vessels have 8,000-TEU or greater capacity.[6] The largest container ships in service in 2005—such as *CSCL Asia* and the P&O Nedlloyd *Mondrian* carry 8,500 TEUs.

Thanks to Captains Jon Harrison and Mark Remijan, I was able to visit *APL China* on one of her stops in the Port of Los Angeles. She has a rating of 4,832 TEUs and is 905 feet long with a beam of 132 feet. Harrison and Remijan alternate as masters of *APL China*, working a shift of 70 days—the time it takes for the vessel to make two round-trips between ports on the West Coast of the United States and ports in Asia.

Harrison and Remijan took me on a tour of the vessel, from the engine room to the bridge. Containers are stored below deck in a series of bays with racks, nine layers high. The containers are attached to each other and to the racks by a latching mechanism at each corner called an *interbox connector* (IBC). On deck are seven layers—the first three layers secured by lashings in addition to the interbox connectors. It takes roughly three days to unload and reload the ship. In the captain's cabin are several computers with satellite links for downloading the latest weather information. On my visit, one computer screen displayed the ship's most recent track across the Pacific—a smooth crossing, no rough weather.

The trend toward larger vessels also applies to passenger and cruise ships—examples being Cunard's *Queen Elizabeth 2*, 963 feet long, and the even newer *Queen Mary 2*, 1,131 feet long. Royal Caribbean Cruises Ltd. and the Carnival Group are reportedly planning new cruise ships that are even larger.

So what do these trends imply for the safety of the world's merchant fleet? Today's vessels are larger and longer and of necessity traverse such hazardous seas as the treacherous waters of the Agulhas Current with greater frequency. Because of design economies, it appears that vessels are less able to withstand the impact of an extreme wave. At the same time, a large part of the global fleet is 20 years old or older. How will new and old vessels fare in encounters with extreme waves? Only time will tell.

Building double-hulled tankers should improve safety, as long as the interior spaces between hulls can be inspected and maintained. Access to some portions is confined and difficult, and if care is not taken, corrosion or structural problems could go undetected.

Heavy seas have been known to shake loose and claim the containers on container ships or damage them. One problem is a phenomenon known as *parametric rolling*, which occurs in heavy seas when the length of the ship is close to the wavelength. Depending on the speed of the vessel relative to the wave, the vessel will sometimes be in a trough and sometimes be supported on a crest. Container ships feature streamlined hulls designed for speed. Because of the load they carry, these ships have a wide flare on the bow and the stern to provide maximum deck surface. Consequently, the hull area presented to the sea varies, depending on whether the vessel is on a crest (where its broad, flat hull gives maximum stability and restoring force) or in a trough (where the restoring force is less). Then, if the vessel starts encountering waves at about twice its natural roll frequency, it will tend to roll. The restoring force brings the vessel back, but on a crest it overcompensates in effect. In the next trough, the restoring force is less, so the vessel rolls even more, and the rolls become larger and larger. It takes a drastic maneuver—an abrupt turn and speed change—to break this vicious cycle of dangerous rolling. The shipmaster must be quick to recognize the problem and take action, or the ship motion can quickly get out of control.[7] Warning systems have been developed to alert the crew to impending problems.

Passenger ships have been disabled when large waves broke windows and flooded the bridge, rendering electronic systems inoperable. Design criteria for critical ship control areas may need to be revisited.

POSTMORTEM: LOSS OF THE *DERBYSHIRE*

Bulk and combination carriers appear to be the most problematic of vessels at risk, hull and hatch cover design being critical for safety. The loss of the *Derbyshire* (discussed in Chapter 3) illustrates the importance of these ship design details for large vessels that may encounter extreme waves. The disaster has been reviewed extensively, and we can hope that if insights gained can provide for safer ship design, the loss of the *Derbyshire* crew and the agony their family members have endured these many years will not have been entirely in vain.

The question is: Should the combination carrier *Derbyshire* have been able to survive an extreme wave incident? In order to answer this question it is necessary to evaluate the strength of the hatch covers, since they are the vulnerable points of this type of vessel. If they fail, and water floods two compartments or more, there is a high likelihood the vessel will sink. Several postmortems of the *Derbyshire* disaster have been made, with the results indicating that the hatch covers would collapse if the vessel was hit by a wave around 47 feet high. Based on estimates made following the storm, Typhoon Orchid produced waves 70 to 85 feet high.[8] These results suggest that the design of the *Derbyshire* was inadequate. The vessel did not have the ability to withstand the storm and therefore was doomed from its onset.

The technical standards related to the design, construction, and surveying of ships are established by organizations known as *classification societies*. Classification originated in England in the 1700s, when marine insurers developed a system for independent inspection of the vessels they intended to insure. Those in excellent shape were classified "A," while other vowels were assigned for less than excellent condition. Pioneered by Lloyds of London, the concept of ship classification was embraced by other seafaring nations—today there are approximately

50 organizations involved. In 1968, seven major classification societies formed the International Association of Classification Societies, which today has 10 members. They are the American Bureau of Shipping, Bureau Veritas (France), China Classification Society, Det Norske Veritas (Norway), Germanischer Lloyd, the Indian Register of Shipping, Korean Register of Shipping, Lloyd's Register (England), Nippon Kaiji Kyokai (Japan), Registro Italiano Navale (Italy), and the Russian Maritime Register of Shipping. The International Association of Classification Societies (IACS) claims that more than 90 percent of the world's cargo-carrying tonnage is covered by the classification design, construction, and periodic inspections carried out in conformance with the rules and standards of its member organizations, including 95 percent of the world fleet of bulk carriers.[9]

Today ships are not assigned grades: they are either "in class" or not. If the ship operator fails to maintain the vessel to the required standards of care, or fails to submit the vessel for timely periodic inspections, classification may be withdrawn. Vessels are subject to an array of standards promulgated by such international conventions as the International Maritime Organization, by the flag state wherein the vessel is registered, and by its classification society. The International Association of Classification Societies supports the member organizations in several ways. It issues resolutions on technical or procedural matters, makes unified interpretations of matters arising from international conventions, and makes technical recommendations to the members. With the exception of procedural requirements, the association's recommendations do not have to be accepted by the members. One of the association's long-term goals is to develop unified standards and procedures for its members and, of course, to increase maritime safety and reliability.

Noting the high rate of loss of bulk carriers, the association and its members initiated investigations and research to improve vessel safety. A series of documents have been developed and made available to member societies and vessel operators.[10] The International Maritime Organization has also issued structural survivability guidelines for new and existing bulk carriers and new requirements for surveying the ships.

The IACS states that the majority of bulk carriers lost were more than 15 years old, were carrying iron ore at the time, and failed as the result of corrosion and cracking of the structure within cargo spaces, and as a result of overstressing by incorrect cargo-loading and cargo-discharging operations.[11] There is no mention of extreme waves or rough seas as a cause of failure. The IACS has issued standard wave data—called IACS Recommendation 34—for use in the design of cargo-carrying vessels in the North Atlantic. Table 1 in that document indicates that most of the waves will have periods of 8 to 14 seconds and significant heights of 36 feet or less. In fact, the table indicates that around 98 percent of the waves to be encountered will be in the range of 6.5 to 36 feet and only 2 percent will be in the range of 39 to 56 feet, the highest waves listed.[12]

In response to growing discontent by ship owners concerned about the fact that ships being built today are less robust, three classification societies announced in 2001 that they would work together to establish common design criteria for standard ship types, beginning with tankers. In 2004, the chairman of the IACS council, Ugo Salerno, issued a letter reporting on the status of common rules for oil tankers and bulk carriers.[13] The letter stated that two project teams—one known as the Joint Tanker Project and the other the Joint Bulker Project—have been formed. The objective of these projects is to develop new ship design rules. Salerno stated that IACS's objective is that the new rules will be adopted and applied uniformly by all IACS members.

As part of the Joint Tanker Project work, a survey was made of defects found in existing tankers. It showed that fractures and cracking accounted for two-thirds of the defects. The new tanker rules are tentatively scheduled for release in April 2006 and will apply to tankers designed and constructed after that date. The design wave loads in the new rules will be based on IACS Recommendation 34, described previously.

In view of the tragic loss of the *Derbyshire* and numerous other vessels, as well as the unacceptably high loss rate among bulk carriers in general, it is timely for ship classification societies and other organizations that establish ship design criteria to change the rules. Obviously, no vessel can be made absolutely safe. Design is always a compromise between safety, economics, and risk. To give an extreme example, a vessel cannot be designed to withstand 300-foot-high waves; it would be impractical to build and not economical to operate. Moreover, the probability of encountering such a wave is virtually nil. On the other hand, there is sufficient evidence to conclude that 66-foot-high waves can be experienced in the 25-year lifetime of oceangoing vessels, and that 98-foot-high waves are less likely but not out of the question. Therefore, a design criterion based on 36-foot-high waves seems inadequate when the risk of losing crew and cargo is considered. IACS Recommendation 34 should be modified so that the *minimum* design wave height is at least 65 feet. The *dynamic* force of wave impacts should also be included in the structural analysis (as opposed to relying on static or quasi-dynamic analyses).

Good vessel maintenance is equally important, particularly to prevent corrosion of hulls and vital structural elements. Saltwater corrosion eats away unprotected steel, reducing its strength—a phenomenon ship surveyors call "wastage." In 1960, the tanker vessel *Pine Ridge* was surveyed and given clearance to sail, even though wastage in key structural members ranged from 25 to 60 percent, meaning that the thickness of vital parts of the ship was only 25 to 60 percent of the original design value. Today, with thinner sections and less steel in modern vessels, careful maintenance and inspection are vital to control wastage. Could a modern vessel with 60 percent wastage survive a gale, let alone a serious storm? In 1960, the tanker *Pine Ridge* could not; she broke in half, and seven crew members, including the master, lost their lives.

In the coming decade the world's shipyards will be busy fulfilling orders for new tankers to replace the old single-hull vessels that must be withdrawn from service, as well as building more bulk carriers, building more and larger container ships, and building even larger passenger vessels. With fewer vessels overall in the merchant fleet, and with replacement vessels being more costly but more economical to operate, there is a benefit to be gained by investing in more robust designs, better safety equipment, more extensive crew training, improved weather instrumentation and weather routing, and conducting inspection and maintenance to rigorous standards. This opportunity to improve the safety and reliability of merchant vessels should not be lost.

11

Davy Jones's Locker

The origin of the term Davy Jones's locker is not known with certainty. One speculation is that Jones ran a waterfront tavern frequented by sailors. He supposedly drugged them and confined them in his ale locker so they could be pressed into service by the British navy. Even if the origin of the term is obscure, the meaning is not. Davy Jones's locker connotes a cold, watery grave, eternal rest at an unknown location at the bottom of the ocean. Many a person has suffered this fate when violent storms and huge waves overtook them. Some were famous, but most were sailors or passengers or coastal dwellers known only to their families and relatives. It is even fair to say that extreme waves and storms have changed the course of history, as in the case of the typhoon—"Divine Winds"—that devastated a Chinese fleet sailing to invade Japan in 1291, or the big waves that destroyed the Spanish Armada in 1588 and ultimately led to the ascendancy of British naval power.

Civilization is in shock over the tragedy of the December 26, 2004, tsunami in Southeast Asia. The magnitude of suffering beggars understanding. Parents who experienced the unspeakable terror of their infants being ripped from their arms by the fury of the waves will live with that horror forever. Families torn apart will always wonder what happened to missing loved ones. As I worked on this book in the days following the disaster, the estimate of deaths, first 20,000, doubled each day, eventually reaching 280,000. Sadly, we recognize that the true toll in human lives will never be known. Scientists will agonize as they ask themselves: "Could this catastrophe have been predicted?" Governments will be challenged by angry citizens: "Could warnings have been given, could aid have arrived faster, could more victims have been pulled from the sea or otherwise rescued?"

The reality is that earthquakes cannot be prevented and some earthquakes will cause tsunami. What is preventable is needless loss of life. The knowledge exists to build earthquake-resistant and flood-resistant buildings; for economic and other reasons it is not always employed. Warning systems can be deployed—satellite communication technology today makes such systems a bargain compared to the cost of just one day of international aid to tsunami victims.

There are conditions besides tsunami that can give rise to extreme waves—waves that can cause appalling property loss and loss of human lives in ships at sea. Extreme waves are likewise not preventable, but again, needless loss of life is preventable through the design of better vessels and improved weather forecasting models. Pressures due to economics and competition need to be examined to see if they are placing ships and crews at excessive risk. The research program currently under way within the European community is a much needed first step. Hopefully, this work will lead to better predictive tools and new design criteria for both seagoing vessels and offshore structures that will better enable them to withstand extreme waves.

We can never quiet the stirrings of Ruau-Moko in the womb of the earth, nor can we still the restless waters raised by Poseidon. But through awareness, and better science and engineering, we could certainly improve the odds for survival. The keys are better design, forecasting and prediction, and warning systems.

DESIGN OF PORT AND COASTAL FACILITIES

Today, more people than ever before live along the shorelines of every country touching upon an ocean. Over the ages, the oceans have risen and receded. Vast parts of the world that are today dry land were once ocean bottoms. Today there is clear evidence that the oceans are advancing; scientists dispute the causes—is it global warming due to human activities, or is it an inevitable cycling of global temperatures? To islanders such as those living on the coral islands of Tuvalu and Kiribati, it does not matter; they just know their land is disappearing. So, it is not just the risk of a tsunami that makes the design of port and coastal facilities so important; if the mean sea level rises, some coastal facilities will have a reduced margin of safety; others will simply disappear.

Just what *is* the margin of safety? Elsewhere I've mentioned that the elevation of the Balboa Peninsula where I live is 10 feet above mean sea level—not much comfort there. The same is true of large sections of the Pacific and Atlantic coasts of the United States, as it is of countries ringing the Indian Ocean and of many Pacific Islands. There are two piers, about a mile apart, that are an integral part of our beachfront community. The southerly one, called the Balboa Pier, has a deck that rises 20 feet above mean sea level, while the Newport Pier, a little farther north, is 22 feet above mean sea level. Both piers have restaurants overlooking the ocean, and during daylight hours visitors or people fishing will be on the piers enjoying the view. Would they be safe in a tsunami?

About 15 miles farther north is the giant port complex of the combined Los Angeles–Long Beach harbors. This seaport is the largest in the United States in terms of container ships; every day dozens of giant vessels enter the port from all over the world to discharge goods or to load materials for shipment overseas. There is a long breakwater that protects the harbor; it is 14 feet above mean sea level.

I met with Tony Gioiello, the chief engineer of the Port of Los Angeles, to discuss port design. He told me that the port had contracted with the U.S. Army Corps of Engineers to construct a large-scale hydraulic model of the port, along with its breakwater, various piers, and other features. This model can be used to study tidal effects within the port, seiching, currents, and any other aspects that might affect its operation.

To comply with California's strict seismic design codes, the cranes and other critical port facilities are designed to withstand a major earthquake without collapse. They successfully withstood large earthquakes, including the disastrous 1994 Northridge earthquake. Gioiello told me that port officials are concerned about the possibility of a tsunami damaging the port or interrupting port traffic and about the resulting economic impact. He sent engineers from the port to inspect damage in Southeast Asia following the tsunami there, and they concluded that from the viewpoint of Port of Los Angeles operations, earthquake damage rather than flooding is the most serious concern.[1]

Ports all over the world are expanding to accommodate larger vessels. The port complex at Los Angeles–Long Beach is no exception. Courtesy of Daniel, Mann, Johnson and Mendenhall and affiliated companies—major participants in the design and construction management of port facilities—I got a firsthand look at the new Pier 400 terminal facilities in the Port of Los Angeles. It is the world's largest proprietary container terminal, served by 12 huge post-Panamax cranes that tower over the giant container ships and transfer containers onto an extensive rail system that then takes them by rail to destinations throughout the United States. The facility is designed to withstand a major earthquake.

Regulations concerning the design of coastal facilities are often the responsibility of several different agencies—there is no single entity. For example, in the United States, state and local codes apply onshore and in the coastal zone, while federal regulations apply outside the state boundaries. A comprehensive new regulation for the design of marine facilities (primarily marine oil terminals and related facilities) was recently developed in California.[2] I reviewed this document; it is comprehensive, incorporates the latest research findings, and would serve as a good model for all types of coastal facilities anywhere in the world. It includes wind and wave loads, earthquake, and tsunami design guidance.

The challenge of protecting beachfront homes and businesses is much more complex. In an effort to counter tsunami damage, Japan has erected tsunami walls at a number of coastal locations, including the city of Tsu on the island of Honshu. Tsu, situated on Ise Bay, about 37 miles south of Nagoya, has been hit by a number of tsunami. Some tsunami walls are nearly 15 feet high, but tsunami have been known to overflow them. For example, the Hokkaido earthquake of July 12, 1993, produced one of the largest tsunami in Japanese history—extremely large waves resulting in a 98-foot-high run-up. The tsunami hit the port town of Aonae on the small island of Okushiri, near the west side of Hokkaido within minutes after the earthquake. The town was surrounded by a 14.75-foot-high tsunami wall. Despite the wall, major damage occurred—including the loss of 340 homes—and although the wall may have blunted the force of the waves to some extent, 114 people died in Aonae alone.[3]

States such as California and Florida have thousands of miles of unprotected coastline, much of it expensive beachfront property. The safe thing to do would be to restrict the beaches to swimmers and sunbathers, and build houses on higher ground. However, this has not happened and is not likely to happen; homeowners instead will play the odds and hope that the "100-year event" does not occur while they own the property.

OFFSHORE STRUCTURE DESIGN

Offshore structures are designed for one of the most hostile environments imaginable. Only structures designed for outer space face more severe material and engineering challenges. In the offshore environment, a structure must withstand not only the daily stress of wind and wave, but the rare extreme event as well as the long-term effect of fatigue and saltwater corrosion.

Offshore structures are predominantly those used by the petroleum industry. They are typically designed to withstand the fatigue due to flexing caused by normal wind, wave, and current forces and to withstand extreme conditions, usually expressed in terms of an event that occurs only once in 100 years. The extreme design criteria differ from area to area—for example, from the Gulf of Mexico to the west coast of Africa or to the North Sea. The northern North Sea provides a good example of one of the more severe environments:[4]

- 100-year extreme wave height: 102 feet
- 1-year summer storm wave height   46 feet

- Wind   80 knots
- Current   2.9 knots

By comparison, in the Gulf of Mexico the 100-year extreme wave is 72 feet.

Offshore platforms are constructed with steel legs or concrete columns, or may float and be held in place by an anchoring system. A common configuration is the steel jacket type, which resembles a bar stool, but typically has eight or more legs. In the North Sea, the typical water depth for such a platform is 328 feet. A number of platforms have been instrumented to measure incident wave heights and periods, movement of the platform decks, stresses, and related information. Measured data can then be compared with the values used to design the platform in an effort to assess the margin of safety.

For example, a metal-jacketed platform in the Frigg field, northern North Sea, was analyzed using several years of recorded data.[5] In a 1981 storm when the significant wave height was 45 feet, the maximum wave height was around 62 feet. This wave caused the deck to move approximately 3.5 inches. When these results were extrapolated to the case of the 100-year wave, the extreme displacements were found to be 90 percent or less of the design values, indicating that the structure would survive the extreme wave unless it was somehow weakened beforehand.

What could weaken the structure? Shifting or settlement of the sea bottom could reduce the strength of the support legs, or corrosion could weaken them. Over time, the constant pounding of the sea could weaken or break some of the numerous cross members and braces of the support structure. Alternatively, marine life growing on the legs and underwater structure could increase its effective mass and the area presented to waves, causing the structure to experience greater forces and impacts than would occur in the clean condition.

SHIP DESIGN

To understand why ships sink, you need to know some basics of ship flotation and stability, and the implications for ship design. Throw a piece of wood in a lake, a stream, or the ocean and it will be seen to roll and bob with the motion of the water. These motions resemble those of an airplane; likewise a ship can move in six different ways: Forward, sideways, vertically (up and down); as well as *roll* sideways, *pitch* up and down, and *yaw* about a vertical axis. Once a vessel has been designed for level floating in calm water, with sufficient *freeboard* (clearance above the waterline), when it is fully loaded and the load properly distributed so the decks stay high and dry, its roll and pitch are the two characteristics of greatest interest.

The weight of a ship and its cargo may be thought of as a force acting downward through its center of gravity. Opposing this downward force is an upward force—the buoyancy force—that keeps the vessel afloat. When the ship is upright and level, the buoyancy force acts through a point called the center of buoyancy that is aligned with the center of gravity. As the ship rolls, the shape of that part of the hull presented to the water changes, and the buoyancy force now pushes upward at a point away from the center of the ship (more toward the side of the hull), causing the ship to return to an upright position. When the ship rolls, the buoyancy force is directed through a point above the center of gravity called the *metacenter*. The *metacentric height* is the distance from the center of gravity to the metacenter. It is usually abbreviated GM.

A vessel with a large metacentric height and weight low in the vessel (a tanker or dry bulk ship) is very stable—that is, it can heel over a great deal and still return to an upright position. Such a vessel is also said to be "stiff," meaning it responds quickly to sea motions and is uncomfortable to passengers. On the other hand, a vessel with a smaller metacentric height (a passenger ship) will be more comfortable to passengers because of its slower response to rolling. This type of vessel is said to be "tender." (Remember reading about the SS *Waratah* elsewhere in this book?)

The cargo in a bulk carrier, container ship, or tanker, when properly stored and evenly distributed, ensures that the vessel rides in the water as the designers intended—not too high, so as to be top heavy, and not too low, so as to be subject to breaking seas. In the empty state, vessels take on ballast to maintain stability.

In the old days of wooden sailing vessels, small stones often made up the ballast. In California's early history, trading vessels would sail up from Mexico to collect cargos of hides and tallow from the cattle ranches and haciendas. A favorite loading spot was Catalina Harbor, on the back side of Catalina Island. The ships would drop anchor there and unload their load of ballast stones at one side of the harbor. Today these stones from old Mexico form a long finger of land that curls out into the harbor in the shape of the letter C. The California Yacht Club (Marina Del Rey, California) has its "Ballast Point Station" there.

Ballasting is very important. Today, most vessels incorporate special tanks into which seawater can be pumped to ballast the vessel. Another approach is to pump water into an empty fuel tank. Some shipmasters are reluctant to do this unless absolutely necessary, because later the tank must be cleaned before it can be refilled with fuel—resulting in a possible delay. On occasion, this hesitancy has led to the loss of the vessel in a storm.

Captain Jerry Fee was the director of ship design for the U.S. Navy and also served as the president of the American Society of Naval Engineers for a number of years.[6] When discussing ship design with Fee, I asked him how he came to be a ship designer. He told me that in addition to being a graduate of the U.S. Naval Academy and spending three years at the Massachusetts Institute of Technology in postgraduate naval engineering training, he spent five years on active duty on destroyers and then a number of years in the Navy's salvage and repair operations. From these experiences, he developed firsthand knowledge of the effects of heavy weather on naval vessels, as the following incident shows.

"In February 1962, I was serving as junior officer of the deck on a Fletcher class destroyer—the USS *Taylor*—DD468," said Fee. "We were involved in patrols and maneuvers about 100 nautical miles off of the east coast of Russia, west of the Aleutian islands and not far from Kamchatka in the Bering Sea. For four days the destroyer wallowed in extremely heavy seas so rough that the ward room was closed and the only meals were sandwiches grabbed at random. The *Taylor* rolled about 40 degrees in 30- to 40-foot-high waves during the storm, on occasion as much as 53 degrees—so much that from the flying bridge

50 feet above the waterline it seemed as though you could reach out and touch the water.

"The U.S. squadron was frequently overflown by Soviet aircraft monitoring us during the storm. On the bridge, officers braced themselves with one foot on the bulkhead (wall) and the other on the deck, due to the angle of the ship. A rogue wave—estimated to be 60 feet high—crashed over the bow of the ship, arched over the entire foredeck, and hit the top of the flying bridge, demolishing its cover. The flying bridge had an aluminum enclosure to protect the crew from the weather. As the roof collapsed around me, tons of water drove me to my knees and I had to grab a stanchion to avoid being washed off the bridge as the water poured off the ship."

There is no substitute for personal experience when it comes to designing ships to withstand heavy weather!

As mentioned above, every vessel has a range of stability that depends on its design and load. If the vessel rolls beyond the range of stability, the vessel will no longer be able to right itself and will capsize. Fee told me that Navy destroyers are designed to be able to recover from extreme rolls approaching 90 degrees. In towing tank tests the Fletcher class (DD445) model recovers from rolls as great as 110 degrees; however, in a real situation, once the roll exceeds 90 degrees, the uptakes (air intake or engine exhausts) are likely to take on water, which would severely inhibit the ship's recovery from a roll.

Fee went on to explain that in addition to designing vessels for stability, ship designers must build into them sufficient strength to withstand the forces imposed by operations, normal seas, and storms. The "backbone" of a ship is its hull, a steel structure not unlike a steel building laid on its side. Key structural elements are defined by the *scantlings*, or design specifications. Hull design is examined from two perspectives: *sagging*, a condition whereby the vessel is supported fore and aft on two large waves and the center sags, and *hogging*, where the vessel is supported by a single large wave in the center and the bow and stern sections are unsupported. The hull is designed and reinforced to withstand the extreme loads imposed by whichever of these two conditions is most extreme. Next, the hull and deck plates are designed to withstand the impact of wave and water. This can take two forms: the deadweight of tons of water that can result from a large wave washing over the deck, and the "slap" force or impact of a fast-moving wave striking the ship. Interestingly, this is more crucial for a large vessel than for a small ship; the small ship will more likely move with the wave, whereas the large vessel, possessing huge inertia, is more likely to be battered as an immovable object by the wave. (See Plates 14 and 15.)

Typical maximum fetch for a local storm over the ocean is 500 nautical miles. If the wind blows for a long enough period of time— say, a day and a half to two days over this distance, waves with a significant wave height of 33 to 49 feet will be produced. Such waves have a wavelength of 300 to 600 feet, the length of a good-sized ship. This is one of the reasons they can be so damaging—if the vessel rides up on the crest of the wave with the bow and stern unsupported. Tankers and bulk carriers (vessels carrying coal, grain, ore, cement, etc.) are susceptible because of the design. If the vessel is partially supported on a large wave, hogging or sagging (bending) failure can occur.[7]

THE LOSS OF THE *MARINE ELECTRIC*

The breakup of the *Derbyshire* in Typhoon Orchid near Okinawa is described in Chapters 3 and 10. Nearly 20 years passed from the time the vessel disappeared until the British government finally conducted a formal investigation into the loss of the ship. The hearings brought a sense of closure to the grieving relatives of the lost crew members, although that was slight consolation given their loss and the length of time that had passed. The two-decade-long ordeal of resolving the *Derbyshire* mystery has had an additional benefit—to focus attention on the appalling loss of bulk carriers and crew. It is unfortunate that this investigation did not take place earlier, because in the interim dozens of bulk carriers continued to fall victim to large waves.

The loss of the *Marine Electric* occurred in stormy seas about 27 nautical miles off the coast of Virginia on Friday, February 11, 1983, during one of the worst storms to hit the East Coast in more than 40 years. The waves were not the highest ever experienced in that area, but at 40 feet they were certainly high enough. In other ways, the *Marine Electric* incident is representative of nearly all the elements involved in ship design and safety at sea in heavy weather.[8]

To start with, the *Marine Electric* was a reconditioned World War II ship. Originally built as a tanker, she was modified in 1961 to be a bulk carrier. This process included adding a new center section and extending the length of the vessel to 605 feet. Her normal run was to carry 24,000 tons of coal from Norfolk, Virginia, and deliver it to the Boston area, where it was used as fuel in electric power plants. The first and perhaps most elementary aspect of safety at sea is "don't overload the vessel." Cargo ships are marked with a set of load lines, called Plimsoll marks that are approved by regulatory bodies and insurance carriers. Plimsoll marks show where the waterline should be when the vessel is fully loaded. The correct level varies with salinity, water temperature, and service conditions. In tropical freshwater the load line (labeled TF) is higher, meaning the ship can ride lower in the water, whereas in cold, dense water, the load line is lower, meaning the ship has to ride higher. An example is the load line labeled WNA, for winter conditions in the North Atlantic. A ship loaded in a freshwater port will magically ride higher in the ocean because of the greater buoyancy of saltwater.

Equally important is the distribution of the load throughout the ship. Obviously, it cannot all be on one side or the vessel will develop a dangerous list in that direction. To prevent this situation from occurring accidentally, cargo ships and tankers have baffles built into their holds or use other methods to keep the cargo from shifting. If the load is concentrated forward and aft, the vessel will have a tendency to "hog" and potentially break apart in the center. Equally problematic is concentrating the load amidships, so that the vessel could sag in the center when supported fore and aft by a large wave. Also important is the condition of hatches and ventilators—anything that could let water into the vessel during heavy seas. Flooding can obviously disrupt the balance of a fully loaded vessel. Pumps capable of pumping out water in the case of a leak somewhere are essential. The *Marine Electric* had a problem in this regard. The deck and many hatch covers were cracked. They had been crudely patched, but despite repeated requests in writing by the crew, they were not repaired.

At midnight on February 10, 1983, *Marine Electric* sailed. Weather advisories were posted with gale warnings. As noted in Chapter 2, this means winds in excess of 34 knots. By the afternoon of the next day, the weather had worsened to a Force 10 gale. Seas were heavier, and occasionally the *Marine Electric* plowed her way through a rogue wave estimated at 60 feet high. It was about this time that the crew spotted the *Theodora*, a 65-foot fishing boat. They passed each other in the storm. A short while later the *Theodora* was heard on the radio calling the coast guard. She was taking on water and in danger of sinking— and did not know her exact position. *Marine Electric* responded to the coast guard with an estimate of the *Theodora*'s last position. At this point, the coast guard asked *Marine Electric* to turn around, locate the other vessel, and stand by until help arrived.

Once, I was sailing downwind from San Miguel Island back to Santa Rosa Island in *Dreams*. The seas were moderate, with waves about 6 feet high. Three of us were on a diving and fishing trip to the Channel Islands. I knew we were heading into rough water, so I'd advised the crew to lash all of the dive gear securely on deck. I happened to look back to see what kind of waves were chasing us when I spotted a black object in the water. It was a dive bag loaded with a regulator, wet suit, dive computer—at least a thousand dollars worth of gear. Retrieving it meant coming about and chasing it before it sank. In this maneuver, timing is important. You want to turn the boat (if possible) in the trough between waves, quickly enough that the next wave does not catch you broadside and roll you. This maneuver successfully executed, the dive bag was pulled aboard a few minutes later with a boat hook just before it sank and was lost forever.

With this experience in mind, you have to admire the nerve of the captain of the *Marine Electric*. Reversing course with a vessel nearly 20 times longer than *Dreams* and in seas five times higher was no easy feat. *Marine Electric* successfully came about and headed back to the distressed vessel. Now *Marine Electric* was taking heavy seas on her stern. She stood by the *Theodora* for several hours until the coast guard arrived and managed to drop some pumps to the *Theodora* so she was able to control flooding and make her way to port. Then *Marine Electric* once again made the turn and resumed her former course, bow into the waves.

By the evening of February 11, *Marine Electric* was in trouble herself. The vessel was no longer plowing through the big waves. She was down at the head, as they say, meaning the bow was riding too low in the water. By 3:00 A.M., the crew realized that *Marine Electric* was not going to make it. They radioed their position to the coast guard and requested immediate help, then changed course to try to make Delaware Bay.

A few minutes later the large vessel was listing heavily. The crew tried to ready the lifeboats. While they were in the process, the boat rolled over and capsized. Some crew jumped; some were tossed into the frigid water. Some were trapped inside the foundering vessel and never had a chance. By luck, one man found a life raft with capacity for 15 men in the violent seas and hung on to it. He was joined by three others. After struggling for minutes the first man succeeded in getting into the raft. He tried to get the others into the raft as it was being pummeled by huge waves. Their hands were too cold to grip. Despite his efforts, he could not pull them in. One by one, they drifted away to certain death. Finally, a coast guard helicopter arrived on the scene; three men survived the ordeal; the rest of the crew—31 sailors—including the captain, were lost at sea.

In the aftermath of the sinking, a marine board of inquiry found that the cause of sinking was "fracture of the vessel." The coast guard's report placed the blame on water leaking in through the cracks in the deck and failed hatch covers.[9] The coast guard report stated that, in accordance with the International Convention on Load Lines of 1966, hatch covers were "to be designed for an assumed load of 358 pounds per square foot with a safety factor of 4.25 on the material ultimate strength." Note that at 64 pounds per cubic foot seawater density, it would take just 5.6 feet of seawater on top of the hatches to reach this loading. There is little doubt that waves much larger than this slammed on top of the hatch covers.

What can we learn from the loss of vessels such as the *Marine Electric* and the *Derbyshire?*

- An obvious conclusion is that design standards need to be reviewed in light of what is currently known concerning the size and occurrence of extreme waves. A wave 50 feet high carries four times as much energy as a wave 25 feet high, and a wave 100 feet high has *16* times the energy. A vessel designed for 25-foot-high seas might not be able to withstand the tremendous forces imposed by extreme waves— forces that can collapse hatch covers, buckle hull plates, or even break the back of the vessel.

• There is no single, internationally accepted standard-making body for ship design, no uniform set of rules. Instead, rules are promulgated by various classification societies for each type of vessel, and each classification society has its own set of rules. Design loadings are generally expressed in engineering terms, and it is not clear what sea state they are based on. From the sample I've been able to review, it appears that a sea state with 35-foot-high waves is the norm.[10] In light of current knowledge, this is inadequate. I would not want to serve on a 900-foot-long vessel built to these standards and routinely traversing the Gulf Stream, or the Agulhas or Kuroshio currents.

• A secondary conclusion is that the masters of vessels designed in accordance with the older standards need to know the capability of their ships and need to have the authority to avoid those waters where there is a higher probability of encountering extreme waves. Requiring a rigid adherence to schedule and route may cause the vessel and crew to be endangered needlessly.

• Finally, if the vessel is not properly maintained, there is no telling what kind of seas she will be able to withstand if put to the ultimate test. Normal practice is dry dock every five years, at which time the vessel is supposed to be inspected for cracks and damage due to corrosion. Older vessels frequently had hull plates that were 1 inch thick. Given today's high-strength steels, the thickness has been reduced in many cases to 0.4 inches. In addition, new epoxy paints provide better corrosion protection.[11] But if defects in the epoxy paint are not detected and repaired, the margin of safety for corrosion damage is less than it used to be.

MARINE WEATHER FORECASTING AND ROUTING

The first line of defense against extreme waves is never to encounter one. While there is no absolute guarantee that this can be done, mariners can improve the odds in their favor by avoiding weather conditions known to give rise to extreme waves and by avoiding those parts of the oceans where their incidence is greatest. Marine weather charts provide sea state conditions (surface wind speed and direction, significant wave heights, swell period and direction, isobar lines, atmospheric pressure highs and lows, locations and movements of developing storms or gales). Refer back to Figure 9 for an example of a wind-wave forecast. Upper air (500-millibar) charts show the direction and strength of the upper air winds that have an important effect on surface conditions. Winds at the 500-millibar level generally blow from west to east. By examining the "troughs and ridges" of upper-air isobars, forecasters can anticipate where surface lows and highs will occur. Other charts show tropical cyclone danger areas, sea surface temperature, and satellite imagery.[12] Weather analyses are issued once or twice per day, and forecasts typically are issued for 24, 48, and 96 hours, although other periods also may be used.

Weather charts and text reports are available to ships at sea via shortwave (single sideband) radio or satellite links. Today, even small cruising yachts can access these reports. For example, on *Dreams*, using a laptop computer, a terminal node controller (a fancy name for a radio-frequency modem, a device that encodes or decodes text or graphics files transmitted by radio), and a high-frequency shortwave radio, I can download the latest marine weather charts. With this capability, a vessel can keep track of weather in its vicinity as well as distant weather trends.

At present, the ability to forecast extreme waves does not exist. Researchers are working to identify sea state conditions that can be correlated with the occurrence of extreme waves. If the work is successful, it may be possible to provide warnings to vessels to avoid areas where such waves could occur. In particular, a parameter called the Benjamin-Feir instability index may correlate with sea conditions where extreme waves are more likely. European weather forecasters are experimenting with this approach.[13] There are global ocean maps that indicate prevailing significant wave heights for each month of the year. This information, combined with knowledge of major currents such as the Agulhas or Gulf Stream, provides an indication of areas where extreme waves are more likely to occur.

While most of the marine weather data are in the public domain and free, there are also commercial services that provide specialized weather services on a subscription or fee basis. Captains Jon Harrison and Mark Remijan demonstrated this type of capability when I visited them on board *APL China*. They have access (via satellite and e-mail) to the latest NOAA and other weather reports to enable them to plot their own routes across the Pacific Ocean, or they can use a weather routing service. Harrison and Remijan said that generally they do their own routing using a program developed by Dr. Henry Chen. The program is very sophisticated and enables the ship master to program acceptance criteria for a route—that is, beam seas no higher than 5 meters, head seas no greater than 6 meters, no rolling greater than 20 degrees, no tropical storm approach (at 35-knot wind speed contour) closer than 50 nautical miles, etc. The program then takes the latest weather data and the ship's parameters and computes a route that satisfies the input criteria or, if this is impossible, reports "no solution." Different routes can be evaluated by simulating against forecast weather conditions.

Subsequently, I visited Dr. Chen, the president of Ocean Systems, Inc., at his office in Alameda, California. He demonstrated how a shipboard computer would download forecast current, wind, and wave information for an ocean area of interest. The program he developed is called Vessel Optimization and Safety System (VOSS). A unique feature is that it has the ship response characteristics (such as fuel consumption versus speed, roll and pitch periods, etc.) generated for the ship's actual loading condition (drafts and GM) and stored in memory. The ship's master can input safe operating envelope parameters unique to the vessel. These can include such parameters as maximum wave, maximum wind speed, maximum roll angle, number of bow slams per hour, number of times green water hits on the deck per hour, and other conditions. With this information, the master can select a destination and the program will define the optimum route that avoids exceeding the safe operating envelope and minimizes the fuel consumption for the desired arrival time. Alternate routes can be evaluated in terms of total fuel consumption, average speed, and estimated time en route so the master can evaluate the consequences of one versus another. Global wind and wave forecast data for VOSS are provided through a partnership with Oceanweather, Inc.[14]

Weather routers provide specialized forecasts for sailors and also will provide weather updates and suggested course changes during a voyage. They can supply a detailed weather outlook for the area to be sailed. For example, in 2001 I sailed in *Dreams* to Isla Guadalupe (Mexico), an island that is 150 nautical miles offshore from Baja California at 29 degrees north latitude. Since my two crew members (Russell Spencer and Erik Oistad) had been with me on a previous attempt when we altered our plans due to bad weather and never made it to the island, Erik suggested that this time we make use of a weather routing service. We were also concerned because two weeks earlier Hurricane Juliette had hammered Cabo San Lucas at the tip of Baja California. We selected weather router Rick Shema and told him our destination and expected departure date. A few days before we were ready to sail, he advised that the weather looked favorable and gave us a report on what we could expect to encounter. Periodically we advised him by e-mail (over the single sideband radio) of actual conditions, and he gave us an updated forecast halfway through the two-week trip. Also, as part of his service, he monitored conditions along our route and was prepared to contact us by e-mail if bad weather was headed our way. We had good weather during the entire trip, but it was reassuring to know that we had recourse to a second opinion if the weather deteriorated.[15]

While preparing this book I called Shema and asked him if he thought it possible to include extreme wave warnings in his forecasts. After some research he said: "It might be possible by incorporating some indices relating wave steepness to the wave spectrum, and then introducing additional wave parameters into wave forecast models, to identify ocean areas with greater probability of extreme wave formation. The marine forecaster routing ships could then alert the vessel to the possible risks and offer alternative routing to avoid high-risk areas. More research is necessary before this can be done routinely. Forecasting the timing and location of individual extreme waves is unlikely due to the randomness and complexity involved in their formation." Knowing Shema, I suspect it is just a matter of time before he'll come up with an extreme wave advisory service for his customers.

EXTREME WAVE PREDICTION

If better vessel design and avoiding hazardous weather conditions are the first weapons in protecting against the dangers of extreme waves, providing warnings when they are about to occur has to be the next most important step we can take. Tsunami warning systems are described in Chapter 6. The technology exists and has been demonstrated; what remains is implementation and public awareness on a global scale.

For other forms of extreme waves, scientists are working to develop ways of predicting what sea conditions could conceivably lead to extreme waves. The work is being pursued on a number of fronts: developing remote sensing technologies that permit the measurement of wave heights and direction using high-frequency, satellite-based radar systems; making improved sea state forecasting models; creating maps or atlases that advise mariners where extreme waves are most likely to occur; developing instrumentation systems including shipborne radar that can measure sea conditions; and carrying out research aimed at improving the design of vessels, offshore structures, and coastal facilities to withstand the impact of extreme waves.

Dreams is equipped with radar having a 24-nautical-mile maximum range. This is very useful when I am attempting to avoid other vessels at night or in fog. One phenomenon that affects radar sensitivity is so-called *backscatter*. Backscatter is defined as radar waves that are reflected to the vessel by things other than the target vessel, including rain or signals reflected by the sea itself. Reflection from the sea surface is called *sea clutter*. Remarkably, scientists have found ways to use satellite radar measurements of sea clutter to detect extreme waves. In another development, wave detection systems using marine x-band radar (radar with a 3- to 6-centimeter wavelength) have been installed on offshore platforms as well as on ships.[16]

Researchers such as Dr. Susanne Lehner and others are working to increase the availability of measured data and to improve wave forecasting algorithms. Radar data have been compared to data obtained by wave-riding buoys that measure wave height and direction, good accuracy being reported. Dr. Lehner and her colleagues have done pioneering work in this field using data collected by synthetic aperture radar on a remote sensing satellite operated by the European Space Agency.[17] Using representative data collected during the summer and fall of 1996, Lehner found a number of waves characterized by extreme height, steepness, and asymmetry (crests higher than trough depth). Near Antarctica, waves as high as 97.7 feet were detected in an area where the sea state consisted of waves with a significant height of 33 feet, so the ratio was 2.9, indicative of rogue waves. Researchers also reconstructed waves in the storm track of Hurricane Fran in the Atlantic; there the waves reached heights of 50 to 60 feet. An extension of this work that Lehner has demonstrated is to provide global maps of the highest individual wave heights and maximum wave steepness; such information, if available routinely on a daily forecast basis, would be a godsend to mariners and represents a very exciting development in this field.

CONCLUSIONS

Humans have lived in a symbiotic relationship with the sea for millennia. In addition to serving as essential sources of food and providing important means of transportation, oceans have an enduring quality—a draw that does not fade with time. Thus it is unlikely that humans will abandon coastal areas because of the remote risk of a tsunami. Memories often fade following a disaster—a generation may pass, and then people move back to risky areas. With the growth of the world population, more and more people are likely live in coastal regions.

Other than building mammoth seawalls, there is little that can be done to protect coastal areas within a few hundred kilometers of a tsunami origin. However, as noted in Chapter 6, tsunami education programs that teach people to evacuate coastal areas immediately on foot when a large earthquake is felt will save lives. Tsunami waves travel too fast to make a warning possible, and even if a warning is given, in densely populated coastal communities, evacuation by automobile on congested roads may become impossible. So tsunami may be a risk that has to be accepted in coastal communities.

Fatalism should not be allowed to preclude the installation of extensive warning systems in coastal areas surrounding all oceans. Early warning systems are cheap compared to the cost of lives lost. The benefit of early warning systems has been amply demonstrated by the success of hurricane warnings and evacuations in the southeastern United States and other areas. Also, work continues on the science of earthquake prediction. While today it is not possible to predict the size and magnitude of impending earthquakes, it is reasonable to hope that at some future time this will be possible. Prediction could give a sufficient lead time to nearby communities if the time frame were measured in hours rather than minutes.

ARE SHIPS' CREWS BEING SENT TO SEA TO DIE NEEDLESSLY?

Likewise, prediction and mapping of extreme wave risk areas as a routine part of weather forecasts will save hundreds of lives every year. With the current trend toward larger vessels and longer voyages, the risk to mariners is increasing and the ability to avoid rogue waves takes on an even greater importance. I get the impression that certain classes of vessels have overemphasized construction economies at the expense of crew safety. In conducting the research for this book, I was shocked at the shipping loss statistics I found—a major vessel, every day or two, somewhere in the world. Ironically, with the environmental sensitivity that exists today in most parts of the world, if an oil tanker spills a few hundred barrels of oil on someone's beach, it is front-page news. But let a 650-foot-long bulk carrier suddenly disappear with 30,000 tons of cargo and its entire crew, and it may only be noted in passing in the newspapers.

For the family members of the crew, it is a different matter, of course. They live with the terrible uncertainty of not knowing what really happened. They may live with this uncertainty for years, as in the case of the *Derbyshire*, until investigators in deep-water submersible vessels finally find the broken remains of the vessel on the ocean bottom and are able to piece together what happened or, as is the case in many such losses (the *Waratah* being a prime example), what happened is never known. There is just the terrible void caused by those who are never seen again.

I wonder if there isn't a moral issue here. How can shipbuilders and owners countenance the thought that their ship crews are being placed in mortal danger by less than adequate ship design or poorly maintained vessels? Are merchant ship crews expendable? Are relatives and family members of lost crew the only people who care?

For more than two centuries the phrase "Davy Jones's locker" has been synonymous with death in the sea. Today, with an improved understanding of extreme waves, we have the potential to ensure that the loss of a vessel due to giant waves becomes a rare event. It is time to take the necessary steps, time to slam the door shut on Davy Jones's locker.

Appendix A

Recent Research on Extreme Wave Models

As awareness of extreme waves has grown, researchers have intensified efforts to understand the physics underlying their formation. If the occurrence of extreme waves can be linked to certain meteorological or sea state conditions, then it might be possible to forecast when they will occur. Such information would be invaluable to mariners and the marine insurance industry.

Much of what we know about extreme waves—indeed, the emphasis on learning more about them—has come from observations by shipmasters who have witnessed encounters between their vessel and an extreme wave or from survivors of vessels that foundered as a result of an extreme wave incident. Those observations are by their nature imprecise; generally they occurred too suddenly or under extremely stressful conditions where exact measurements were impossible.

In Chapter 8, I grouped extreme waves by probable cause, in the following order:

- Strong currents
- Storms
- Continental shelves and shallow seas

- Constructive interference (superposition)

To these we may add several additional possible mechanisms for extreme wave formation that are the subject of current research:

- Nonlinear effects
- Spatial or temporal focusing effects
- Multidirectional effects
- Modulation and resonance

The first category, strong currents, has indisputably been the source of extreme waves. When swell or storm waves encounter a fast-moving opposing current such as the Agulhas Current or the Gulf Stream, they tend to "pile up" as their velocity is reduced. Professor Chris Garrett pointed out that a wave with a phase speed of c meters per second can be stopped by an opposing current of only $1/4$ c.[1] When this happens, steep, high waves result, proceeded or followed by deep troughs. Thus, the evidence is clear that the probability of encountering an extreme wave is greater under these conditions and a prudent mariner should avoid this situation if possible.

The increase in wave size as a function of wind velocity, fetch, and wind duration is a well-known phenomenon in storms. There are correlations that provide estimates of the significant wave heights under varying storm conditions, but none that predict the random occurrence of extreme waves. This suggests that some additional mechanism, yet to be fully understood, is at work. It would be useful to know if extreme wave formation is governed by a threshold effect; in other words, do seas have to build to a certain point before extreme waves are produced? Or is it purely a statistical effect?

The evidence seems to indicate the latter possibility, because many mariners (myself included) recall sailing in relatively calm seas where the significant wave height was a few feet, but suddenly a wave two to three times as high struck the vessel.

Likewise, shallow water, bottom effects, and refraction have the effect of slowing waves and causing wave heights to increase. Areas where there is a sharp transition in sea depth are potential danger zones in rough seas. The question here is: Are there certain sea or wind conditions that combine to cause an extreme wave and how can they be anticipated?

Constructive interference (sometimes called superposition) is the fourth category mentioned in Chapter 8. Here I include incidents (to the best of my knowledge) in which an extreme wave struck a vessel in the absence of the other conditions described above. Superposition is a well-known phenomenon, observable in many areas of physics, so its existence is not in doubt. The relevant research question is whether or not it is capable of causing waves that are 2.2 to 2.4 times as high as the significant wave. From the Rayleigh distribution, the answer would seem to be yes. (A factor of two seems obvious.)

Turning to new areas of research, attempts to model nonlinear wave effects may shed light on how superposition can produce extreme waves. Nonlinearity makes mathematical modeling much more difficult, so scientists and engineers always try to develop linear models first and then turn to nonlinear models as a last resort. Some researchers go so far as to say that if rogue waves exist, they must inherently be nonlinear. Some of the features of extreme waves—their steepness and the shape of the wave crest—are modeled more accurately when second-order terms are included.

Another interesting new approach is based on spatial and temporal focusing. Spatial focusing is another way of describing what happens when waves are refracted by the ocean bottom in coastal waters or by current gradients.[2] Temporal focusing may result when waves disperse. Some wave groups may contract to a few wavelengths and then combine with others to produce short groups of very large waves.[3] Some theorists believe that nonlinear focusing may allow a wave to "borrow" energy from its neighbors, becoming as much as 4.5 to 5 times greater than the average wave height.[4]

Multidirectional and multidimensional effects are also being studied to see if they can cause extreme waves. The idea is to investigate whether extreme waves result from wave trains interacting at an angle or from three-dimensional interactions—effects that would not be modeled by a one-dimensional analysis. When I spoke with Dr. Susanne Lehner, she indicated that waves from crossing seas arriving from two different storms can add and build to an extreme size due to the continuous input of energy. Also, the Benjamin-Feir index, introduced in Chapter 11, measures a phenomenon called the Benjamin-Feir instability. It is defined as the ratio of the mean square slope of the frequency spectrum peak to its normalized width. Under the right conditions, instability causes the wave train to break up into periodic groups. Within each group a further focusing takes place, producing a very large and steep wave having a height roughly three times the initial height of the wave train.[5]

Finally, other research is directed at seeing whether frequency or amplitude modulation could be responsible for extreme waves. Or is it possible that certain wave periods and frequencies will resonate with a given sea state condition to create extreme waves? The analogy to this is easily demonstrated in a bathtub. A bathtub, or a harbor for that matter, has a series of resonant frequencies. If you take a piece of wood (or possibly just your hand) and get the water sloshing back and forth at just the right period, a large wave will occur.

Appendix B

Units of Measure and Conversion Factors

Multiply:	By:	To get:
Bars	10^5	Pascals (newton per square meter)
Bars	14.5	Pounds per square inch
Barrels (U.S.)	42	Gallons
Barrels (U.S.)	0.119	Cubic meters
Cubic feet	0.0283	Cubic meters
Feet	0.305	Meters
Feet	12	Inches
Grams per cubic centimeter	1.0	Metric tons per cubic meter
Inches	2.54	Centimeters
Kilograms per square meter	0.205	Pounds per square foot
Kilometers	0.621	Statute miles
Kilometers	0.540	Nautical miles
Knots	0.514	Meters per second
Knots	1.151	Miles per hour
Meters	3.281	Feet

Multiply:	By:	To get:
Meters	39.37	Inches
Metric tons	2,005	Pounds
Miles	1.609	Kilometers
Miles	5,280	Feet
Miles per hour	0.869	Knots
Millibars	0.0295	Inches of mercury at 32 degrees Fahrenheit
Nautical miles	1.852	Kilometers
Nautical miles	1.151	Statute miles
Pounds (force)	4.448	Newtons
Pounds	0.454	Kilograms
Pounds per square foot	47.9	Pascals (newtons per square meter)
Pounds per square inch	6,890	Pascals
Pounds per cubic foot	0.01602	Metric tons per cubic meter
Register tons	100	Cubic feet
Register tons	2.832	Cubic meters
Short tons (2,000 pounds)	0.907	Metric tons

Appendix C

Glossary of Special Terms

beam. The width of a vessel at its widest point; also the side of a ship.

Beaufort wind scale. A visual description of the sea state used to estimate wind speeds and wave heights.

broach. To cause a vessel to veer and turn its beam (side) to the wind and waves.

capsize. To roll a vessel on its beam and to cause it to overturn.

coaming. A raised wall or border around the edge of a ship, its hatches, or other openings to prevent water from entering.

dead reckoning. The determination of a vessel's position based on course and distance run.

deadweight tonnage. The carrying capacity of a vessel. Originally expressed in long tons (2,240 pounds), today the standard is metric tons.

displacement tonnage. The actual weight of a vessel. Originally expressed in long tons (2,240 pounds), today the standard is metric tons. It is numerically equal to the weight of the water displaced by the vessel (i.e., the water beneath the hull).

draft (or draught). The distance from the surface of the water to the lowest part of the vessel beneath the surface.

drift. The speed of a current in knots.

fetch. The distance of open ocean over which the wind has been blowing.

FPSO. Floating production, storage, and offloading system. A special tanker designed to be moored in an offshore oil field.

freeboard. The distance from the waterline to the lowest part of the deck of a vessel.

GM. An abbreviation for the distance from the center of gravity of a ship to its metacenter; a measure of a vessel's stability.

gross tonnage. The total enclosed space or internal capacity of a vessel calculated in tons of 100 cubic feet (2.83 cubic meters) each. This is derived from old merchant ship traditions where 1 ton of merchandise on average occupied 100 cubic feet. Under the new regulations of the International Maritime Organization, it is now expressed in cubic meters.

heaving to. A maneuver to protect a boat and crew in rough seas. For a powered vessel, it usually means to head into oncoming seas, at a slight angle to the waves, with just enough speed to maintain steerage way through the water. For a sailboat, after reducing sail, the jib is backed (so the wind is against the sail, rather than pushing it), and the main sail or storm trysail is set to barely maintain forward motion. The rudder is then brought to the leeward side of the vessel and locked there. When the sails are adjusted properly the boat is essentially stationary, making very slow forward progress while also moving sideways.

Mayday. The international radiotelephone distress signal. It is derived from the French *m' aider* ("help me"). Before the advent of voice radio communications, in Morse code the distress signal was SOS, signifying "Save Our Souls." It is transmitted or signaled as . . . - - - . . ., or dot dot dot, dash dash dash, dot dot dot.

metacentric height. Literally, the height of the metacenter of a floating vessel above its center of gravity. Its significance is that it is a measure of ship stability.

net (registered) tonnage. A measurement of the earning power of a vessel carrying cargo, equal to the gross tonnage minus spaces on the vessel that cannot carry cargo, such as the engine room. This is often used as the basis for assessing fees on the vessel.

OBO. Literally, ore-bulk-oil, meaning a combination carrier, a bulk carrier that can carry various types of cargo. The *Derbyshire* is an example.

pitchpole. When a vessel flips end over end, usually because of sailing too fast down the face of a high, steep wave, the bow becomes buried in the trough and the breaking wave carries the stern over.

Plimsoll marks. Marks painted on the side of a vessel that indicate how deeply it can be submerged under various sea conditions.

rig. On a sailboat, the spars (masts), sails, and standing rigging.

rode. A line connecting an anchor to a vessel; it can be rope or chain.

Ro-Ro carrier. Ro-Ro stands for "roll on–roll off" and signifies a vessel that carries cargo that is driven on and off the vessel.

run-up. Referring to a tsunami, the maximum elevation reached by the wave on land.

scantlings. A set of standard dimensions for ship design.

set. The direction toward which a current is flowing.

significant wave height. The average height of the highest one-third of a group of waves.

soliton. A solitary wave. Solitons can propagate on the surface or below.

SOS. See Mayday.

TEU: Twenty-foot equivalent unit. Refers to the original standard size of containers used on container ships. Today, 40-foot containers are common; one of these would equal 2 TEUs.

Endnotes

INTRODUCTION

1 Bascom (1980), 153, points out that 9 percent of all ships ever built in the 6,000 years of ocean navigation were wooden sailing ships, with steel, engine-driven vessels being a development of the last century. He estimates that 3,000 vessels were wrecked and 1,000 vessels foundered per year during the 1700s and 1800s—a fantastic number.

2 Mayday is the international radiotelephone distress signal. See Appendix C for the definition of this and other specialized terms. It is equivalent to SOS transmitted by Morse code.

3 At the publisher's insistence, I used the British/American system of units in this book, rather than the universally accepted SI units. A table of conversion factors is found in Appendix B.

4 "Lloyd's List" has been published continuously since the early 1700s when marine insurance underwriters gathered in Lloyd's coffeehouse in London. It describes ship movements from various ports around the world and also lists ship casualties and losses.

5 Bascom (1980), 153, 158.

6 J. C. Chapman and R. Adams (1984), "Structural Design of Mono Hull Ships," 57, in Douglas Faulkner, et al., eds. (1984), *The Role of Design, Inspection, and Redundancy in Marine Structural Reliability*, Proceedings of an International Symposium, November 14-16, 1983, Committee on Marine Structures, National Research Council, Washington, DC: National Academy Press.

7 The information may be found on *www.cargolaw.com*, the web site of Los

Angeles-based Countryman and McDaniel, a law office specializing in international trade and maritime transport.

8 Richard Monden and Dieter Stockman (2004), "ISL Market Analysis 2004: World Shipbuilding and Maritime Casualties," 4-5, in *Shipping Statistics and Market Review,* March 2005, Institute of Shipping Economics and Logistics (ISL).

9 Committee on Tank Vessel Design, National Academy of Sciences (1991), 14. *Tanker Spills—Prevention by Design.* National Research Council, National Academy of Sciences, Washington, DC: National Academy Press.

10 Charles E. Herdendorf and Judy Conrad (1991), "Hurricane Gold: Part I—The Loss," 4-17, *Mariner's Weather Log,* Vol. 35, No. 3, 5-10. See also Thompson (1998), 58.

11 Anonymous (1857), 281, *Frank Leslie's Illustrated Newspaper.* New York. Also mentioned in Thompson (1998), 58.

CHAPTER 1

1 *Dreams* is a Hans Christian sailboat, an ocean cruising cutter with a full keel. She was launched in 1987. I purchased her in 1993 for the purpose of undertaking longer voyages not possible in *Karess.*

2 Zebrowski (1997), 150; Kinsman (1965), 11.

3 Hendrickson (1984), 122.

4 Homer (1996), 161-162.

5 Herodotus (1972), 283.

6 For details concerning fifteenth-century Chinese navigators, refer to Parry (1981), 39-40; Bergreen (2004), 232-238; and Levathes (1994).

7 Columbus's remarkable journey is described in a number of excellent books. See Dyson (1991), 64. See also Taviani (1989); Morison (1978), 351-548; and Cummins (1992), 79-133.

8 Among the best descriptions of Magellan's epic voyage is Bergreen (2004), 132-171, 391-392.

9 See Kinsman (1965). His book, with its eloquent and thought-provoking notes and asides, is fascinating reading.

10 Oceanographer Michel Ochi provides an excellent overview of these new developments. See Ochi (1998).

CHAPTER 2

1 Price (1990), 64.

2 Two books provide a description of the race. See Knecht (2001), 306, for Ellison's remark. Mundle (1999), 116, 164-169, 170-175, presents the stories of several other competitors.

3 Professor P. N. Joubert, University of Melbourne, Victoria, Australia, personal communication, March 2006. Professor Joubert has sailed many times in the Sydney-Hobart on his boat *Kingurra.* For additional details concerning this inci-

dent, see P. N. Joubert (2006), "Some Remarks of the 1998 Sydney-Hobart Race," i-x, Transactions in *Proceedings of the Royal Society of Victoria*, Vol. 117, No. 2.

4 Bruce Johnson (2000), "Capsize Resistance and Survivability When Smaller Vessels Encounter Extreme Waves," 48-49 in M. Olagnon and G. A. Athanassoulis, eds. (2001), *Rogue Waves 2000*, Proceedings of a workshop organized by Ifremer, November 29-30, 2000, Brest, France.

5 Peter Lewis, personal communication, July 5, 2005.

6 Van Dorn (1974), 69-71.

7 Preston and Preston (2004), 5.

8 Based on Van Dorn (1974), 75.

9 This summary description of currents is based on Bowditch (2002), 434-438.

10 To a small boat, hitting a partially submerged container at night in the middle of the Pacific is one of a yachtsman's worst nightmares—equivalent to the *Titanic* hitting an iceberg.

11 Curtis C. Ebbesmeyer and W. James Ingraham, Jr. (1992), "Shoe Spill in the North Pacific," *Eos,* Vol. 73, No. 34, August 25, 361-368.

12 One thousand leagues is 5,556 kilometers, or 3,452 miles.

13 Sverdrup et al. (2005), 236-238.

14 Hendrickson (1984), 135. In the early days of sailing, sailors used a log line to measure speed. The line consisted of a cord knotted at equal intervals of 15.5 meters (51 feet). At the end of the line was a triangular piece of wood (the log) weighted at one edge, the line attached so that the log floated vertically, almost like a kite. When thrown overboard from the stern of the vessel and the line allowed to flow out freely, the log would remain more or less stationary in the water. The navigator would count the number of knots passing through his hand while watching an hourglass timed for half a minute. At the end of this interval, the number of knots that had passed was equal to the speed of the vessel in nautical miles per hour—thus the term knot; since a nautical mile equals 6,076 feet, 120 knots is equal to 1 mile; this is also the number of half-minutes (120) in an hour, so the number of knots that pass over the stern in a half-minute is equal to speed in knots.

15 Massel (1996), 21-50. Theoretician Stanislaw Massel provides a detailed summary for anyone wanting to probe deeper into the mathematics of this subject. In summary, when wind first flows over calm water, atmospheric pressure fluctuations are advected (meaning horizontal air movement that causes changes in its properties) over the sea's surface, creating a resonant effect that causes small wavelets to form.

16 Sverdrup et al. (2005), 254.

17 For conversion purposes, 1 bar = 0.987 atmosphere = 29.53 inches of mercury at 32 degrees Fahrenheit. Usual units are inches of mercury, millibars, or hectopascals (hPa), which are numerically equal to millibars, so the former term is used throughout this text because it is more familiar to most readers.

18 Farrington (1996), 116-122, 239-240.

19 Lochhaas (2003), 246-248.

20 *Newsweek,* "Solo Against the Sea," April 4, 2005, 38-39.

CHAPTER 3

1 A synopsis of the *Derbyshire* loss and subsequent investigation can be found at *http://www.nautical-heritage.org.uk/derbyshire.html*. The multiagency (government and Department of Defense, United States and Canada) Ship Structure Committee is a cooperative entity dedicated to research in the realm of ship structures, materials, and vessel performance; *http://www.shipstructure.org*. I have relied on its investigation of the *Derbyshire* loss as described in Daniel Tarman and Edgar Heitmann, "Case Study II: *Derbyshire*-Loss of a Bulk Carrier," Washington, DC: Ship Structure Committee.

2 After Muga (1984), 159.

3 For more information on the probability of large waves, see Ochi (1998), 58; Young (1999), 25-26; and Bruce J. Muga, "Statistical Descriptions of Ocean Waves," Chapter 6 in Wilson (1984), 159-161.

4 Young (1999), 27. Note: Be warned that in spite of this, the largest wave can be the third, the ninth, the tenth, etc.

5 Van Dorn (1974), 192-199.

6 Ochi (1998), 255-280.

7 Young (1994), 38-40.

8 Bowditch (2002), 443-449.

9 If you are uncomfortable with visualizing volumes of water, think of people coming down an escalator and imagine that those on the bottom are slow to exit. Pile-up!

10 Bowditch (2002), 449.

11 Adapted from Kinsman (1965), 11.

12 Bascom (1980), 95-111, Chapter 5, "Tides and Seiches." This small book has one of the clearest descriptions of tidal behavior that I've encountered and is a remarkable book in other respects. It has been out of print for some time.

13 Jane Hollingsworth (1989), "The Chicago Seiches," *Mariners Weather Log*, Vol. 3, No. 2, Spring, 16-19.

14 LeBlond (1978), 512.

15 This account is based on Beach (1966). Beach, a captain in the U.S. Navy, is perhaps best known for his classic book on submarines, *Run Silent, Run Deep*. His father, Edward L. Beach, Sr., was captain of the *Memphis*. See particularly pp. 52 and 96 for wave heights.

16 This estimate is based on the sailor's rule of thumb that at sea level you can see an object 3 meters (10 feet) high at a distance of 3.6 nautical miles, or 30 meters (100 feet) high at 11.4 nautical miles, the limitation being the curvature of the earth. At the level of the *Memphis* bridge (assumed to be 50 feet above sea level), a 23-meter (75-foot) high wave could be seen at a distance of around 18 nautical miles. (This assumes perfect visibility, no haze or fog, good binoculars, etc.) See Chapman (1976), 603, for table of visibility of objects at sea.

17 Dr. George Pararas-Carayannis, personal communication, May 1, 2005. See also George Pararas-Carayannis (2005), "The Loss of the USS *Memphis* on 29 August 1916—Was a Tsunami Responsible?" Available at *http://drgeorgepc.com/ LossUSSMemphis1.hmtl*. Accessed July 2005. This well-documented study demon-

strates that it was most likely swell from hurricanes rather than a tsunami that wrecked the *Memphis*.

CHAPTER 4

1 Rousmaniere (1983), 101-105.

2 Ron Holle, meteorologist, personal communication, August 2005. I thank Mr. Holle for reviewing the weather descriptions and clarifying key points for me. See also Chapman (1976), 241-242.

3 Typical maximum fetch for a local storm over the ocean is 500 nautical miles. This will produce waves with a significant wave height of 10 to 15 meters, crest to trough, if the wind blows for a long enough time, say a day and a half to two days. So Brad is not exaggerating; if anything, with wind speeds in excess of 70 knots, waves were higher than his estimate. See Chapter 2 and Sverdrup et al. (2005), 254.

4 Walker (2001), 111-260. In his book, Walker describes a half-dozen incidents involving rescues in terrible storms. In one instance, a coast guard pilot allowed his helicopter to descend a fraction too low and it was snatched from the air by a giant wave and lost with the entire crew.

5 Attributed to Patrick Etheridge, who served in the U.S. Life-Saving Service, an agency established by Congress in 1878. It was later merged with the U.S. Revenue Cutter Service to form the U.S. Coast Guard in 1915. See *www.uscg.mil/hq/g-cp/history/faqs/LSSmotto.html*. Accessed October 10, 2005.

6 Trumbull (1942), 205.

7 Couper (1983), 57.

8 See Sheets and Williams (2001), 1-2, 32-33, 191-193, 285-286.

9 Bowditch (2002), 503-505; also U.S National Weather Service, National Hurricane Center, *http://www.nhc.noaa.gov*, which is the source for the Saffir-Simpson Hurricane Scale, statistical data on hurricanes, Table 3 from Eric S. Blake et al., *The Deadliest, Costliest, and Most Intense United States Tropical Cyclones from 1851 to 2004 (and other frequently requested hurricane facts)*, NOAA Technical Memorandum NWS TPC-4, National Weather Service, National Hurricane Center, Tropical Prediction Center, updated August 2005; also see "Frequently Asked Questions."

10 Newport Harbor Museum (1992), *The Hurricane of 1939*, video documentary.

11 Ochi (2003), vii.

12 This example is based on a storm history in Bowditch (2002), 508-510.

13 Young (1999), 150.

14 Underwater instruments measured a 27.5-meter (91-foot) high wave during the passage of Hurricane Ivan on September 15, 2004. See D. W. Wang, W. J. Teague, et al. (2005), "Extreme Seas Under the Eye of Ivan," *Joint Assembly of the American Geophysical Union*, May 23-27; Sid Perkins (2005), ". . . and churn up big waves, too," *Science News*, June 11, Vol. 167, No. 24, 382.

15 Young (1999), 144-160.

16 Krieger (2002).

17 Sverdrup et al. (2005), 204-205.

18 Simpson and Riehl (1981), 19. See also Hendrickson (1984), 308.

19 Larson (1999), 200-203.

20 See Sheets and Williams (2001), 203-221. Readers interested in additional information are advised to consult the National Hurricane Center web site for an overview of the most recent models used for forecasting. The web site also provides references and links to more detailed technical literature for the National Hurricane Center web site; *www.nhc.noaa.gov.*

21 Carrier (2001).

22 Ochi (2003), 141.

CHAPTER 5

1 Warshaw (2003), 619-621.

2 Wave heights are reported in different ways. On the Pacific coast it is the conventional trough-to-crest distance as viewed from the front of the wave. In Hawaii it is the height viewed from the back of the wave, since large waves are associated with reef breaks. Thus, a 21-meter (6.6-foot) wave in Hawaii is 3 to 4 meters (9.8 to 13.1 feet) anywhere else.

3 Warshaw (2003), 644.

4 Jenkins (1999), 2-16.

5 For the Atlantic and Pacific oceans bordering the United States, these forecasts may be accessed on the Internet by going to the National Weather Service home page at *www.nws.noaa.gov* and then clicking on "marine" and finally "weather charts." Weather service meteorologists prepare the forecasts by amassing weather data from weather ships, weather buoys, volunteer ship weather reports, and satellite data and then using computer models to predict future weather conditions. The forecasts are updated several times per day and are issued for 24, 48, 72, and 96 hours. The wind-wave charts show the significant wave heights H_s and display wind arrows that show the direction and speed of the wind. A different set of charts shows the direction and period (in seconds) of the dominant waves. Surface charts provide a compilation of major weather information, including notation of the locations of storms, gales, tropical storms, or hurricanes; atmospheric pressure in millibars; and locations and direction of movements for cold and warm fronts. Other charts show where tropical cyclone danger areas are located and provide sea surface temperatures as well as satellite images of various ocean areas.

6 David Reyes (2005), "Surf's Up and So Is Chance of Flooding," *Los Angeles Times*, March 10, B3.

7 A U.S. Navy web site, *https://www.fnmoc.navy.mil/PUBLIC/WAM,* commonly referred to as the "WAM" site, features colored maps of the North and South Atlantic oceans, the North and South Pacific oceans, and the Indian Ocean, showing wave directions and significant wave heights. The color codes range from dark blue for 0 to 1 meter (0 to 3 foot) wave heights, up to dark brown 14.6 meters (48 foot) wave heights. The forecasts for each region extend out six days in 12-hour intervals, so the movements of large waves can be anticipated. The National Oceanic and Atmospheric Administration, as well as Canada, the Scripps Institution of Oceanography, and other entities operate a series of buoys in the Northwest Pacific

and California coastal areas. These also can be accessed using the Internet. For example, *www.ndbc.noaa.gov* will take you to the locations of moored buoys scattered around the Pacific, as well as to the locations of drifting buoys. To obtain real-time data from a buoy, click on its image on the map. To see an example of a local conditions report, go to *http://facs.scripps.edu/surf/images*, which will lead you to a series of coastal buoys in the Southern California bight. In California, as in Hawaii, Australia, Brazil, Tahiti, and other international surfing spots, there are commercial services that provide surf forecasts. An example is *www.stormsurf.com.*

8 Ge Chen and Jun Ma (2002), "Identification of Swell Zones in the Ocean: A Remote Sensing Approach," 946-948 in IEEE (2002), *Remote Sensing: Integrating Our Views of the Planet,* International Geoscience and Remote Sensing Symposium 2002, 24th Canadian Symposium on Remote Sensing, Toronto, June 24-28, Piscataway, NJ: The Institute of Electrical and Electronic Engineers.

9 Winchester (2004), 277.

CHAPTER 6

1 Bryant (2001), 4. Note that Bryant's book is aptly subtitled "The Underrated Hazard." After the events of December 26, 2004, in Southeast Asia, we can see that the hazard should no longer be underrated. The book is a clear, comprehensive, and excellent resource. Bryant states (p. 21) that there have been 462,597 deaths (1997 data) in the last 2,000 years.

2 Bryant (2001), xxiv, 84-120.

3 Myles (1985) describes the Ras Shamra and Santorini Island events: 154, 170-183; Zebrowski (1997), 7-13.

4 Elias Antar (1971), "Earthquake!" *Saudi Aramco World,* May/June, Vol. 22, No. 3, 5.

5 For information on Japanese tsunami, see Myles (1985), 97-109, 117-125, and Bryant (2001), 21, 49.

6 Preston (2004), 220; Myles (1985), 117-125; Marx (1983), 377-379.

7 Hendrickson (1984), 307, "Memoirs of Capt. L. G. Billings"; Myles (1985), 40-43.

8 Bryant (2001), 21.

9 James F. Lander and Patricia A. Lockridge (1989), 94-96. *United States Tsunamis (Including United States Possessions 1690-1988).* National Geophysical Data Center, Publication 41-2. Boulder, CO: U.S. Department of Commerce; Myles (1985), 47-53; Bryant (2001), 49.

10 Bryant (2001), 51; Sverdrup et al. (2005), 263.

11 Dudley and Lee (1998), 327.

12 Lander and Lockridge (1989), op. cit., 97.

13 See Lander and Lockridge (1989), op. cit.; Bryant (2001), 152-156.

14 Jerry L. Coffman and Carl A. von Hake, eds. (1973), 108-110, *Earthquake History of the United States,* Publication 41-1, Rev. ed. Washington, DC: U.S. Department of Commerce, National Oceanic and Atmospheric Administration; Hendrickson (1984), 308; Bryant (2001), 156.

15 Coffman and von Hake, et al. (1973), loc. cit.; Lander and Lockridge (1989), op. cit. 97-101; Myles (1985), 147, 148 (footnote 28) reports that geologists use the term "swash" to describe such a wave; Hugh Owens (2004), 52. "The Lituya Legacy." *Cruising World*, October, 50-52.

16 Charles L. Mader (1999), "Modeling the 1958 Lituya Bay Mega-Tsunami," *Science of Tsunami Hazards*, Vol. 17, 57-67; George Pararas-Carayannis (1999), "Analysis of Mechanism of Tsunami Generation in Lituya Bay," *Science of Tsunami Hazards*, Vol. 17, 193-206.

17 Bryant (2001), 14.

18 Winchester (2004) is the classic reference, see 240-258; Sverdrup et al. (2005), 263; Prager (2000), 105.

19 N. N. Ambraseys (1962), "Data for the Investigation of the Seismic Sea Waves in the Eastern Mediterranean," *Bulletin of the Seismological Society of America*, 895-913.

CHAPTER 7

1 Jim Gower (2005), "*Jason 1* Detects the 26 December 2004 Tsunami," *EOS Transactions AGU*, Vol. 86, No. 4, 37. Available at *http//www.agu.org/pubs/crosssref/2005/2005 EO040002.shtml*. Accessed October 2, 2005.

2 Jose Borrero, et al. (2005), "Could It Happen Here?" *Civil Engineering*, April, 54-65, 133.

3 J. C. Carracedo (1994), "The Canary Islands: An Example of Structural Control on the Growth of Large Oceanic-Island Volcanoes," *Journal of Volcanology and Geothermal Research*, Vol. 60, 225-241.

CHAPTER 8

1 World Meteorological Organization (1988), 2-4. *Guide to Wave Analysis and Forecasting*. WMO-702. Geneva: World Meteorological Organization.

2 Kinsman (1965), 19.

3 Price (1960), 43.

4 Lewis (1979).

5 Thomas (1997), 81.

6 Lewis (1979), 200-203.

7 This was accomplished under the auspices of the Polynesian Voyaging Society, formed in 1973 by Ben Finney, Herb Kane, and Tommy Holmes. See *http://pvs.kcc.hawaii.edu*. See also Finney (1994).

8 Robert Irving (1993), "Solitary Waves," *Mariners Weather Log*, Vol. 37, No. 4, Fall 1993, 20-23; P. Taylor and C. Swan (2001), "New Waves, Solitons and Spreading," in Olagnon and Athanassoulis (2001), op. cit. 137-142.

9 Flayhart (2003), 69-78.

10 Two excellent references on heavy weather sailing are Pardey (1996) and Coles (1996).

11 This account is extracted from an account compiled by Alf Cook as reported on his web site *www.shipsoflongago.co.uk* and used with his permission. In my correspondence with Mr. Cook, he vividly recalled the excitement generated by Captain Carlsen's determined efforts to save his ship. In addition, I reviewed the official U.S. Coast Guard Marine Board of Investigation report on the *Flying Enterprise* foundering, dated February 26, 1952. This report can be found by accessing *http:// www.uscg.mil/hq* and searching the Marine Board of Investigation data base. See also Butler (1974), 168-169, 183.

12 Lochhaas (2003), 102-107; Aebi (1989), 292-293.

CHAPTER 9

1 British Broadcasting Corporation (2002).

2 Apollonius Rhodius (1998), 39-40.

3 Ochi (1998), 253; and Douglas Faulkner (2001), "Rogue Waves—Defining Their Characteristics for Marine Design" in Olagnon and Athanassoulis (2001), op. cit., 6.

4 British Broadcasting Corporation (2002). See transcript, *www.BBC.co.uk*, quoting Dr. Marten Grundlingh, Council for Scientific and Industrial Research, South Africa; Sverdrup et al. (2005), 255.

5 Goss and Behe (1994), 187. Besides Goss and Behe's account, interested readers can refer to the Board of Trade Court of Inquiry Report No. 7419 on SS *Waratah*, March 17, 1911, London, or J. G. Lockhart (1925), *Mysteries of the Sea: A Book of Strange Tales*, London: Phillip Allen Co.

6 For photographs and details go to the National Underwater & Marine Agency (South Africa) web site at *www.numa.co.za*. NUMA is a nonprofit organization founded by Clive Cussler and supported by his book royalties, for the purpose of preserving maritime history. It has supported explorations at 60-plus historical wrecks. The home page is *www.numa.net*.

7 Captain Jeffrey W. Monroe (1993), "Ship Handling in Heavy Seas," *Mariners Weather Log*, Vol. 37, No. 4, Fall, 10; Committee on Tank Vessel Design (1991), op. cit., 16-17; Bascom (1980), 64.

8 Bascom (1980), 64; British Broadcasting Corporation (2002), 4.

9 Edward J. Barr (1994), "Freak Wave on a Submarine," *Mariners Weather Log*, Vol. 38, No. 4, 34-35.

10 British Broadcasting Corporation (2002).

11 Bascom (1980), 59-60. See also Jerome W. Nickerson (1993), "Freak Waves," *Mariners Weather Log*, Vol. 37, No. 4, Fall, 15.

12 Bascom (1980), 60; Sverre Haver (2001), "Evidences of the Existence of Freak Waves," in Olagnon and Athanassoulis eds. (2001), op. cit., 138.

13 Junger (1997), 71.

14 See *www.ocens.com* for details.

15 For example, Charles Memminger, "Student Earns A-Plus in Sea Ordeal," *Honolulu Star Bulletin*, January 28, 2005, E1; "Semester at Sea Cut Short by 50-Foot Wave," *Los Angeles Times*, January 28, 2005, A17; "Students on Battered Ship Reach Shore," *Los Angeles Times*, February 1, 2005, A14; Kelsey Fronk, "Semester at Sea

Endures Massive Wave in Pacific," *Daily Aztec* (San Diego State University), February 14, 2005.

16 See Bascom (1980), 60; Captain Jeffrey W. Monroe (1993), "Ship Handling in Heavy Seas," *Mariners Weather Log*, Vol. 37, No. 4, Fall, 12.

17 Miles Lawrence (1995) "Preliminary Report, Hurricane Luis, 27 August-11 September, 1995," U.S. National Hurricane Center, *www.nhc.noaa.gov*.

18 Naval History Division (1979), 322. *Dictionary of American Naval Fighting Ships*, Vol. V (letters N through Q). Washington, DC: U.S. Navy Department. See also Kotsch and Henderson (1984), 343, 351.

19 RPMs means engine revolutions per minute. On large ships under normal cruising conditions the vessel speed is often measured in terms of the engine RPMs.

20 Butler (1974), 198-211.

21 I am grateful for the assistance of Chris Chabot, who produced a documentary about the loss of the *Fitzgerald*, and to Fred Shannon, who has carried out extensive research on the sinking of the vessel, for sharing their insights with me. See also Hugh E. Bishop (2001), *The Night the Fitz Went Down*, Duluth, MN: Lake Superior Port Cities, Inc.

22 Thomas Ray (1965), "A History of Texas Towers in Air Defense, 1952-1964," Historical Document 29. Washington, DC: Air Defense Command.

23 Bitner-Gregersen, Elzbieta (2002), 97. "Extreme Wave Crest and Sea State Duration," Appendix B4, A. D. Jenkins et al., *Research Report No. 138*. Bergen: Norwegian Meteorological Institute.

24 Ernie Barker (1998), "Rogue Wave" (unpublished manuscript), 1-2. A single-handed sailor's journal of wave adventures in the Tasman Sea.

25 MaxWave Project (2003), Research Project No. EVK:3-2000-00544. Bergen: Commission of the European Communities. Accessed 11/5/04. Available at *http://w3gkss.de/projects/maxwave*, 2005. Includes scope of work, scientific and technical objectives, work plan, selected technical reports, and a description of the tasks planned in this ambitious and important new research program aimed at improving our knowledge of extreme waves. See "Scientific/Technical Objectives and Innovation," p. 3.

26 Massel (1996), 382-383.

27 Jim Gower and David Jones (1994), "Canadian West Coast Giant Waves," *Mariners Weather Log*, Vol. 38, No. 2, Spring, pp. 4-8.

28 Sverdrup et al. (2005), 254-255. See also Naval History Division (1979), op. cit., 23-24.

29 Ritchie (1996), 176.

30 MaxWave Project (2003), op. cit. See "Scientific/Technical Objectives and Innovation."

CHAPTER 10

1 Committee on Tank Vessel Design (1991), op. cit., 30.

2 Christel Heideloff and Richard Monden (2005), "ISL Market Analysis 2005: World Merchant Fleet, OECD Shipping and Shipbuilding," in *Shipping Statistics*

and Market Review, January/February, Institute of Shipping Economics and Logistics.

3 Ibid., "ISL Market Analysis 2005: Tanker Fleet Development," 1-5.

4 In addition to Captain Jannsen, I am grateful for the hospitality of Captain Giuseppe Mazzoleni, master of the *Cygnus Voyager.*

5 Tarman and Heitmann (no date), op. cit., 3.

6 Heideloff and Monden (2005), op. cit., 2-3.

7 Vadim Belenky (2004), "Demystifying Parametric Roll," *Surveyor*, Fall, 26-29.

8 Tarman and Heitmann (no date), op. cit., 1-18.

9 The International Association of Classification Societies web site (*www.iacs.org.uk*) is the source of this information.

10 Ibid., Recommendation No. 46, *Bulk Carriers—Guidance on Bulk Cargo Loading and Discharging to Reduce the Likelihood of Over-Stressing the Hull Structure* (1997); Recommendation No. 76, *IACS Guidelines for Surveys, Assessment and Repair of Hull Structure—Bulk Carriers;* Revision 2 (2004); Recommendation No. 15, *Care and Survey of Hatch Covers of Dry Cargo Ships—Guidance to Owners;* Revision 2 (1997).

11 Ibid. (1997), *Bulk Carriers—Guidance on Bulk Cargo Loading and Discharging to Reduce the Likelihood of Over-Stressing the Hull Structure.*

12 Ibid. (2001), Recommendation No. 34, *Standard Wave Data*, 2.

13 The International Association of Classification Societies, letter to shipping and shipbuilding associations, etc., from Ugo Salerno. Subject: *IACS Common Rules for Oil Tankers and Bulk Carriers.* Genoa: May 17, 2004.

CHAPTER 11

1 Mr. Tony Gioiello, chief engineer, Port of Los Angeles, personal communication, May 17, 2005. In addition to meeting with me, Mr. Gioiello granted me access to the port's library. I am very grateful to him for his assistance.

2 California State Lands Commission (2005), "Marine Oil Terminal Engineering and Maintenance Standards (MOTEMS)." *California Code of Regulations*, Chapter 31F of Div. 1-11, Title 24, Part 2, Vol. 1, January 19.

3 Eddie Bernard et al. (1993), "Tsunami Devastates Japanese Coastal Region," (Hokkaido Tsunami Survey Group). Seattle, WA: Pacific Marine Environmental Laboratory. This report and photographs of the tsunami wall and tsunami damage can be found at *www.pmel.noaa.gov/tsunami/okushiri.*

4 J. M. Huslid et al. (1983), "Alternate Deep Water Concepts for Northern North Sea Extreme Conditions," 18-49, in Chryssotomos Chryssostomidis and Jerome J. Connor, eds. (1983), *Behaviour of Off-Shore Structures, Proceedings of the Third International Conference*, Vol. 1. New York: Hemisphere.

5 J. Thebault et al. (1985), "In-Service Response Analysis of Two Fixed Offshore Platforms," 123-131 in J. A. Battjes ed. (1985), *Behaviour of Offshore Structures—Proceedings of the 4th International Conference on Behaviour of Offshore Structures* (BOSS '85). Amsterdam: Elsevier.

6 Captain Jerry Fee, U.S. Navy (retired), personal communication,. Captain Fee spent many hours with me, both personally and subsequently by e-mail and telephone, locating information and answers for me when he did not have the answer himself. I am very grateful for his assistance.

7 Sverdrup et al. (2005), 254-255.

8 Frump (2001), 30-31, 45-46.

9 U.S. Coast Guard (1984), "SS Marine Electric, O.N. 245675, Capsizing and Sinking in the Atlantic Ocean on 12 February 1983 with Multiple Loss of Life," *Marine Casualty Report,* No. 16732/0001 HQS 83, Washington, DC.

10 The International Association of Classification Societies has published "Recommendation No. 34: Standard Wave Data," dated November 2001. This recommendation applies to the North Atlantic. The table of sea states shows the most probable waves as being 10.5 meters (34.4 feet) or less. The maximum height listed in the table is 16.5 meters (54.4 feet). According to the table, only 60 to 70 waves out of 100,000 would be 10.5 meters, and only 40 to 50 waves out of 100,000 would be between 10.5 and 16.5 meters (34.4 to 54.4 feet). In a North Atlantic voyage lasting 10 days, 75,000 11.5-second-period waves would be encountered (about 313 waves per hour). On the voyage, the table indicates the possibility that as many as 30 to 40 waves could be in the 10.5- to 16.5-meter (34.4- to 54.4-foot) height range.

11 Captain Jan Jannsen, personal communication, July 8, 2005.

12 An excellent explanation of marine weather charts may be found in Carr (1999).

13 Peter Janssen (2004), "Towards Freak-Wave Prediction Over the Global Oceans," 24-27. European Center for Medium-Range Weather Forecasts, *ECMWF Newsletter,* No. 100, Spring.

14 Henry Chen, Vincent Cardone, and Peter Lacey (1998), "Use of Operation Support Information Technology to Increase Ship Safety and Efficiency," *SNAME Transactions,* Vol. 106, 105-127.

15 A number of weather routing services exist. Rick Shema, *www.weatherguy. com,* (808) 254-2525, is a former U.S. Navy meteorologist who specializes in routing services for racers and cruisers worldwide, with clients in all the major ocean and sea basins. Several others that have provided routing services for everything from commercial shipping to yacht races, including America's Cup, are Weather Routing Inc., *www.wriwx.com,* (518) 798-1110; Commander's Weather, *www.commandersweather.com,* (603) 882-6789, used by many of the Around the World Alone sailors including Brad Van Liew; Skip Gallimore uses OCENS Grib Explorer and weather-on-demand services, *www.ocens.com,* (206) 878-8270.

16 A German research organization developed WaMoS II, a system that uses marine x-band radar to determine the two-dimensional wave spectrum by analyzing the radar images that are scattered back from the sea's surface. Then, using the measured spectrum, a computer program calculates the significant wave height, the period of the peak wave, its wavelength, and its direction. The system uses a dedicated minicomputer that can be connected to an existing shipboard radar system. Wave data are thus available immediately onboard the vessel and show the wave environment for a distance of around 3 nautical miles from the vessel. The WaMoS II system has been installed on both offshore platforms and vessels, and a number

of tests have been made to compare measurement results obtained by it with data obtained by floating buoys and laser wave height measurements. To see an example, go to *www.oceanwaves.org* and click on "real time data."

17 See L. R. Wyatt and J. J. Green, "The Availability and Accuracy of HF Radar Wave Measurements," Vol. 1, 515-517, IEEE (2002); S. Lehner, J. Schulz-Stellenfleth, and A. Niedermeir, "Detection of Extreme Waves Using Synthetic Aperture Radar Images," Vol. 3, 1893-1895, IEEE, (2002); D. Hoja, J. Schulz-Stellenfleth, S. Lehner, and T. Konig, "Global Analysis of Ocean Wave Systems from SAR Wave Mode Data," Vol. 2, 934-936, IEEE (2002).

APPENDIX A

1 Professor Chris Garrett, University of Victoria, British Columbia, personal communication with Craig B. Smith, October 5, 2005.

2 Kristian B. Dysthe (2000), "Modeling a Rogue Wave—Speculations or a Realistic Possibility?" in Olagnon and Athanassoulis (2001), op. cit. 255-264.

3 Efim Pelinovski et al. (2000), "Nonlinear Wave Focusing as a Mechanism of the Freak Wave Generation in the Ocean," in Olagnon and Athanassoulis (2001), op. cit., 193-204.

4 Peter Janssen, "Nonlinear four-wave interaction and freak waves," in Müller (2005), 85-90.

5 Kristian B. Dysthe (2000), "Modeling a Rogue Wave—Speculations or a Realistic Possibility?" in Olagnon and Athanassoulis (2001), op. cit., 255-264; also in the same reference: Miguel Onorato et al., "Occurrence of Freak Waves from Envelope Equations in Random Ocean Wave Simulations," 181, and Efim Pelinovski et al., "Nonlinear Wave Focusing as a Mechanism of the Freak Wave Generation in the Ocean," 193-204.

Annotated Bibliography

Aebi, Tania (with Bernadette Brennan). 1989. *Maiden Voyage*. New York: Ballantine Books. Aebi left New York at age 18, returning two and one-half years later as the youngest woman to sail around the world alone. A fascinating, readable account of a great voyage.

Bascom, Willard. 1980. *Waves and Beaches*. New York: Anchor Press. A classic book, very well written by the man responsible for the Mohole Project.

Beach, Edmund E. 1966. *The Wreck of the Memphis*. New York: Holt, Reinhart and Winston. Story of an armored cruiser driven aground in a shallow harbor by large waves.

Bergreen, Laurence. 2004. *Over the Edge of the World*. New York: Perennial. The story of Magellan's historic circumnavigation.

Bowditch, Nathaniel. 2000. *The American Practical Navigator—An Epitome of Navigation, 2002 Bicentennial Edition*. Bethesda, MD: U.S. Government National Imagery and Mapping Agency. The classic "bible" of sailors and mariners.

British Broadcasting Corporation. 2002. "Rogue Waves." Transcript of a BBC program on freak waves. First aired on BBC TWO, November 14, 2002. An interesting narrative by people interviewed, including the captain of *Queen Elizabeth 2*.

Bryant, Edward. 2001. *Tsunami—The Underrated Hazard*. Cambridge, UK: Cambridge University Press. Bryant is the first author to raise the question of a much greater tsunami risk in a documented, thorough manner. The best book on tsunami there is. His book proved to be prophetic.

Burgess, Robert F. 1970. *Sinkings, Salvages and Shipwrecks*. New York: American Heritage Press. The story of salvage efforts at Port Royal, Jamaica.

Butler, Hal. 1974. *Abandon Ship*. Chicago: Henry Regnery Co. Accounts of the sinking of the *Carl Bradley*, the *Edmund E. Fitzgerald*, and the *Flying Enterprise*.

Carr, Michael W. 1999. *Weather Predicting Simplified.* New York: International Marine. An excellent handbook on weather for the sailor or mariner.

Carrier, Jim. 2001. *The Ship and the Storm—Hurricane Mitch and the Loss of the Fantome.* New York: International Marine/McGraw-Hill. Hurricane (Mitch) that seemed to stalk a modern tall ship, the *Fantome.*

Chapman, Charles F. 1976. *Piloting, Seamanship and Small Boat Handling,* 52nd ed. New York: The Hearst Corporation, Motor Boating and Sailing Division. The bible of all sailors and boat captains.

Coles, K. Adlard (revised by Peter Bruce). 1996. *Heavy Weather Sailing.* Camden, ME: International Marine. A classic book about yacht handling in gales.

Couper, Alistair, ed. 1983. *The Times Atlas of the Oceans.* New York: Van Nostrand Rheinhold.

Cummins, John. 1978. *The Voyage of Christopher Columbus—Columbus' Own Journal of Discovery Newly Restored and Translated.* New York: St. Martin's Press. An annotated version of Columbus' log from his first trip to America.

Dudley, Walter C., and Min Lee. 1998. *Tsunami!* 2nd ed. A Latitude 20 Book. Honolulu: University of Hawaii Press. History of tsunamis with emphasis on the Pacific region, Alaska, and Hawaii. Description of tsunami warning systems and the Pacific Tsunami Warning Center.

Dyson, John. 1991. *Columbus for Gold, God and Glory.* New York: Simon & Schuster. History of Columbus and his discovery of the New World.

Farrington, Tony. 1996. *Rescue in the Pacific—A True Story of Disaster and Survival in a Force 12 Storm.* Camden, ME: International Marine. The story of a weather bomb and its unexpected impact on a sailing regatta from New Zealand to the island of Tonga.

Finney, Ben. 1994. *Voyage of Rediscovery—A Cultural Odyssey Through Polynesia.* Berkeley: University of California Press. Thorough treatment of ancient Polynesian sailing and navigation techniques, along with a description of the *Hokule'a*'s various trips.

Flayhart, William Henry III. 2003. *Perils of the Atlantic—Steamship Disasters, 1850 to Present.* New York: W. W. Norton. Story of the SS *Pennsylvania* and other steamship disasters.

Frazier, Kendrick. 1979. *The Violent Face of Nature—Severe Phenomena and Natural Disasters.* New York: William Morrow. Discusses storms, earthquakes, flooding, and so on, along with forecasting and warning systems, disaster preparation, and response.

Frump, Robert. 2001. *Until the Sea Shall Free Them—Life, Death and Survival in the Merchant Marine.* New York: Doubleday. Detailed account of the *Marine Electric* disaster.

Goss, Michael, and George Behe. 1994. *Lost At Sea—Ghost Ships and Other Mysteries.* Amherst, New York: Prometheus Books. Discussion of the SS *Waratah* incident.

Hendrickson, Robert. 1984. *The Ocean Almanac.* New York: Doubleday. A compendium of miscellaneous facts about ships and the oceans.

Herodotus. 1972. *The Histories.* Translated by Aubrey de Selincourt. London: Penguin Books. A famous Greek tourist tells of his travels in the ancient world.

Homer. 1996. *The Odyssey*. Translated by Robert Fagles. New York: Penguin Putnam. A modern translation of Homer's epic poem concerning Odysseus' 10-year wanderings after the Trojan War.

Jenkins, Bruce. 1999. *North Shore Chronicles—Big-Wave Surfing in Hawaii*. Rev. ed. Berkeley, CA: Frog Ltd. Some people even go looking for extreme waves; this narrative tells the story of the world's most accomplished surfers who have had intimate contact with waves large enough to crush modern oceangoing vessels.

Junger, Sebastian. 1997. *The Perfect Storm—A True Story of Men Against the Sea*. New York: W. W. Norton. The classic story of man trying to survive at sea in the presence of giant waves, with an explanation of one set of conditions that can unexpectedly produce extreme wave conditions.

Kane, Herb K. 1976. *Voyage—The Discovery of Hawaii*. Honolulu: Island Heritage Books. Interesting history of ancient Polynesian navigation by a man who participated in the design, construction, and sailing of the *Hokule'a* from Hawaii to Tahiti.

Kinsman, Blair. 1965. *Wind Waves—Their Generation and Propagation on the Ocean Surface*. Englewood Cliffs, NJ: Prentice-Hall. A classic treatise on wind, waves, and how waves are formed—some dated material, but extremely well written and entertaining.

Knecht, G. Bruce. 2001. *The Proving Ground*. New York: Warner Books. Gripping account of the disastrous 1998 Sydney-Hobart race.

Kotsch, Rear Admiral William J., and Richard Henderson. 1984. *The Heavy Weather Guide*. 2nd ed. Annapolis, MD: Naval Institute Press. A valuable reference book for merchant vessels encountering gales and heavy weather.

Krieger, Michael. 2002. *All the Men in the Sea: The Untold Story of One of the Greatest Rescues in History*. New York: The Free Press. Hundreds of oil field workers are on a barge in the Gulf of Mexico when it is hit by Hurricane Roxanne in 1995.

Larson, Erik. 1999. *Isaac's Storm—A Man, a Time, and the Deadliest Hurricane in History*. New York: Crown Publishers. The chilling story of the destruction of Galveston, Texas, by storm surge and large waves.

LeBlond, Paul H., and Lawrence A. Mysak. 1978. *Waves in the Ocean*. New York: Elsevier. An outdated but classic review of wave theory.

Levathes, Louise. 1994. *When China Ruled the Seas*. New York: Simon and Schuster. The story of early Chinese navigators.

Lewis, David. 1979. *We the Navigators—The Ancient Art of Land Finding in the Pacific*. Honolulu: University Press of Hawaii. Detailed description of ancient Polynesian navigation methods by a man who studied under one of the best surviving navigators.

Lochhaas, Tom, ed. 2003. *Intrepid Voyagers—Stories of the World's Most Adventurous Sailors*. New York: International Marine/McGraw-Hill. A series of accounts of sailors who have sailed around the world alone, crossed the oceans in the smallest boats, or otherwise experienced the sea and its greatest challenges.

Marriott, John. 1987. *Disaster at Sea*. London: Ian Allan. Accounts of various shipwrecks, including a description of the grounding of the *Memphis*.

Marx, Robert F. 1983. *Shipwrecks in the Americas*. New York: Bonanza Books. A history of many shipwrecks in the Americas, with a description of how wrecks are located and a section detailing the exploration of Port Royal, Jamaica.

Massel, Stanislaw R. 1996. *Ocean Surface Waves: Their Physics and Prediction.* Singapore: World Scientific. A very comprehensive review of all of the mathematical theories used to model wind-wave interaction.

Morison, Samuel Eliot. 1978. *The Great Explorers.* Oxford, UK: Oxford University Press. A history of Columbus and other early European explorers and the discovery of America.

Mundle, Rob. 1999. *Fatal Storm—The Inside Story of the Tragic Sydney-Hobart Race.* Camden, ME: International Marine/McGraw Hill. Another account of the Sydney-Hobart race in 1998 as it was hit by a "weather bomb" with devastating results.

Myles, Douglas. 1985. *The Great Wave.* New York: McGraw-Hill. Tsunami incidents.

Naval History Division. 1979. *Dictionary of American Naval Fighting Ships.* Vol. V (letters N through Q). Washington, DC: U.S. Navy Department. Battleship *Pittsburgh* incident in 1945 typhoon.

Ochi, Michel K. 1998. *Ocean Waves—The Stochastic Approach.* Cambridge, UK: Cambridge University Press. A summary of modern wave science and the current methods of analyzing and predicting wave behavior.

———. 2003. *Hurricane-Generated Seas.* New York: Elsevier. A complete treatment of how hurricanes generate extreme waves.

Owens, Hugh. 2004. "The Lituya Legacy." *Cruising World,* October, 50-52. A little-known but huge tsunami caused by an earthquake-induced landslide—remarkable in that there were surviving eyewitnesses.

Pardey, Lin, and Larry Pardey. 1996. *Storm Tactics—Modern Methods of Heaving-to for Survival in Extreme Conditions.* Arcata, CA: Paradise Cay Publications. A valuable handbook for the small-boat sailor planning to venture offshore.

Parry, John H. 1981. *The Discovery of the Sea.* Berkeley: University of California Press. Description of the earliest sailors, including a history of the Chinese navigators.

Prager, Ellen J. 2000. *The Oceans.* New York: McGraw-Hill. A beautiful, almost poetic, treatment of oceans and oceanography.

Preston, Diana, and Michael Preston. 2004. *A Pirate of Exquisite Mind—Explorer, Naturalist and Buccaneer: The Life of William Dampier.* New York: Walker and Company. Describes Dampier's circumnavigations and many contributions to our knowledge of the oceans.

Price, A. Grenfell, ed. 1990. *The Explorations of Captain James Cook in the Pacific—As Told by Selections of His Own Journal 1768-1779.* New York: Dover Publications. Cook's discoveries in his own words.

Rhodius, Apollonius. 1998. Translated by Richard Hunter. *Jason and the Golden Fleece—The Argonautica.* Oxford, UK: Oxford University Press. Retelling of the ancient Greek legend of Jason and his quest for the Golden Fleece that would entitle him to rule in his own right.

Ritchie, David. 1996. *Shipwrecks—An Encyclopedia of the World's Worst Disasters at Sea.* New York: Checkmark Books. This book has capsule descriptions of hundreds of sea disasters listed in alphabetical order, including a number of those cited in this book such as *Poet, Grand Zenith,* and so on.

Rousmaniere, John. 1983. *The Annapolis Book of Seamanship.* New York: Simon and Schuster. A standard guide for sailors in the handling of small craft in rough weather.

Science News. 2005. Vol. 167, No. 24, June 11, 382. Underwater rogue wave measurements off Florida during Hurricane Ivan.

Shaw, David. 2005. "Badly Caught Out, with Tragic Consequences." *Cruising World*, July 18. Rogue wave causes 45-foot Hardin cutter-ketch to founder.

Sheets, Bob, and Jack William. 2001. *Hurricane Watch—Forecasting the Deadliest Storms on Earth*. New York: Vintage Books. A highly recommended book on how hurricanes are formed and describing the history of hurricane forecasting.

Simpson, Robert H., and Herbert Riehl. 1981. *The Hurricane and Its Impact*. Baton Rouge: Louisiana State University Press. A classic treatment of hurricanes by one of the men responsible for the Saffir-Simpson Hurricane Scale.

Sverdrup, Keith A., Alyn C. Duxbury, and Alison B. Duxbury. 2005. *An Introduction to the World's Oceans*. 8th ed. New York: McGraw Hill. A first-rate book for an overall treatment of oceanography.

Taviani, Paolo Emilio. 1989. *Columbus, The Great Adventure—His Life, His Times, and His Voyage*. Translated by Luciano E. Farina and Marc A. Beckwith. New York: Crown Publishers/Orion Books. Review of Columbus and the discovery of America.

Thomas, Steve. 1997. *The Last Navigator*. Camden, ME: International Marine. Story about techniques of ancient Polynesian navigators.

Thompson, Dick. 2004. "Rogue Waves Revealed." *BoatUS Magazine*, November, 34-35. A summary of some of the new findings on extreme waves.

———. 2005. "Roughed Up by Rogue Waves." *BoatUS Magazine*, September, 26-27. Accounts of recent boat sinkings.

Thompson, Tommy. 1998. *America's Lost Treasure*. New York: Atlantic Monthly Press. The amazing story of the discovery and recovery of cargo from the *Central America*, also known as the "Ship of Gold," carrying a cargo of gold bullion and the gold rush miners who had found gold and then lost it in a violent storm in 1857.

Trumbull, Robert. 1942. *The Raft*. Camden, NJ: Henry Holt. The inspiring story of a naval air crew that crash landed in the Pacific during World War II and survived by ingenuity in a small raft.

Van Dorn, William G. 1977. *Oceanography and Seamanship*. New York: Dodd, Mead and Company. Another excellent book on oceans and seamanship. A newer edition of this book is available.

Walker, Spike. 2001. *Coming Back Alive—The True Story of the Most Harrowing Search and Rescue Mission Ever Attempted on Alaska's High Seas*. New York: St. Martin's Press. The story of the U.S. Coast Guard rescue of the crew of the fishing vessel *La Conte*.

Warshaw, Matt. 2003. *The Encyclopedia of Surfing*. New York: Harcourt. Everything you could possibly want to know about surfing may be found here.

Wilson, James F., ed. 1984. *Dynamics of Offshore Structures*. New York: John Wiley & Sons. Engineering methods for the dynamic analysis of offshore structures.

Winchester, Simon. 2004. *Krakatoa—The Day the World Exploded*. New York: Perennial Books. The dramatic story of the island that blew up and the devastating tsunami that followed.

Young, Ian R. 1999. *Wind Generated Ocean Waves*. Oxford, UK: Elsevier. A comprehensive review of wind-wave analytical methods and theory, including wave measurements.

Zebrowski, Ernest, Jr. 1997. *Perils of a Restless Planet—Scientific Perspectives on Natural Disasters*. London: Cambridge University Press. A review of natural disasters (storms, earthquakes, tsunami, volcanoes, asteroid impacts) and their potential or actual impact on life on earth.

Permissions and Credits

I am grateful to the following publishers for permission to reproduce material: Oxford University Press, for the quotation from *Jason and the Golden Fleece: The Argonautica*, Richard Hunter (translator), Oxford World's Classics Series, Oxford University Press, 1998, pp. 39-40; Dover Publications, for the quote from *Explorations of Captain James Cook in the Pacific as Told by Selections of His Own Journals*, edited by Grenfell Price, 1990, p. 64; and Penguin Group (USA) for the quote from "Book 5: Odysseus-Nymph and Shipwreck," pp. 161-162, *The Odyssey* by Homer, Robert Fagles (translator), copyright © 1996 by Robert Fagles. Used by permission of Viking Penguin, a division of Penguin Group (USA).

PLATES

(1) Erik Akiskalian; (2) United States Navy; (3) Richard Bennett; (4) Richard Bennett; (5A) David Groverman; (5B) John Lockwood; (6) Chris Garrett; (7) United States Navy; (8) United States Navy; (9) Henry Helbush; (10) Hellmut Issels / Peter Carrette Icon Images; (11) Eugene Kim/Faye Wachs; (12) Koos van der Lende; (13) John

Lockwood; (14) Krister Brzezinski, *Mariners Weather Log*; (15) United States National Oceanic and Atmospheric Administration, National Weather Service Historic photos.

FIGURES

All figures by Kurt Mueller. Original source for all figures is Craig B. Smith, except as noted: (4) adopted from Prager (2000), p. 88; (8) adopted from Muga, Chapter 6 in Wilson (1984), p. 159 and Young (1999), p. 27; (9) National Oceanic and Atmospheric Administration-National Weather Service; (10) adopted from Young (1999), p. 29: (11) adopted from Kinsman (1965), p. 11: (12) adopted from Young (1999), p. 150; (13) adopted from Bryant (2001) p. 26; (15) adopted from Bryant (2001), p. 153; (18) adopted from Bitner-Gregersen (2002), p. 97.

Acknowledgments

As in the case of all my books, I am indebted to many people for help and encouragement. First, to my business partner, sailing associate, and lifelong friend, Russell Spencer, my thanks for suggesting the topic and encouraging me along the way. Russell, Erik Oistad, Captain Al Gravallese (U.S. Navy, retired), and Jay Winter served as an informal advisory committee and were instrumental in assisting me with many important contacts in the navy, merchant marine, and oceanography fields.

To the many outstanding captains, sailors, and crew who kindly took time to share their expertise and knowledge with me, I offer my most profound thanks: Captain Karl Adams; Captain Jerry Aspland, president, California Maritime Academy (retired); Captain Ernie Barker; Admiral Joe Barth, U.S. Navy (retired); Matthew G. Brown, ChevronTexaco Shipping Company LLC; Captain Bent Christiansen, chief port pilot, Port of Los Angeles; Admiral Bill Cross, U.S. Navy (retired); Captain Scott Culver, Foss Maritime; Bob Degnan, ChevronTexaco; Grant Donesley, voyage manager, ChevronTexaco Shipping Company LLC; Captain Jerry Fee, U.S. Navy (retired); Captain Skip Gallimore, Skipper Keith Garrison, Captain Jon Harrison, American Ship Management, Lieutenant Ray Holdsworth, U.S. Navy (retired); Captain Jan Jannsen, ChevronTexaco Shipping Company LLC; Captain Giuseppe Mazzoleni, Chevron Transport Corporation Ltd.; Captain J. Michael Miller, ChevronTexaco Shipping Company LLC; Captain Mark D. Remijan, American Ship Management; Bill Watkins, vice commodore, California Yacht Club; and Tod and Linda White, S/V *Seascape.*

Eric Akiskalian and Dan Moore told me about the latest developments in tow-in surfing and close encounters with giant waves, while Kent A. Smith provided background material on surfing in general. My colleagues at Daniel, Mann, Johnson and Mendenhall and affiliated companies made port visits possible and provided background information; I particularly thank Clyde Garrison, Greg Hess, and Mike Gasparro, as well as Ms. Behjat Zanjani, president, IEM Corp.

Other technical specialists were generous with their help, including Professor George Pararas-Carayannis, former director of the International Tsunami Information Center (under the auspices of the United Nations Educational, Scientific, and Cultural Organization–Intergovernmental Oceanographic Commission) and chief scientist for various missions sponsored by the United Nations Development Program, Dr. Henry Chen, president, Ocean Systems, Inc.; Alf Cook, Merchant Marine (retired); Hany Elwany, Coastal Environments, Inc.; Gordon Fulton, president, Concept Marine Associates, Inc.; Antonio Gioiello, chief harbor engineer, Port of Los Angeles; Professor Peter Joubert, University of Melbourne; Marion Lyman-Mersereau, Punahou School, Honolulu; John McLaurin, Pacific Merchant Shipping Association; Thomas Russell, general counsel, Port of Los Angeles; Rick Shema, *weatherguy.com*; Dr. Ray Schmitt, Ocean Studies Board, National Academy of Sciences; and Grant Stewart, American Ship Management.

Tom Cain, my master's swimming program colleague and skilled sailor, provided many useful leads on rogue waves. He and Mitzi Wells read early drafts of the manuscript and gave much helpful advice regarding portions that would either overwhelm or bore the general reader. I am grateful to Ronald Holle, meteorologist, for reviewing weather-related material; Professor Chris Garrett, School of Earth and Ocean Sciences, University of Victoria, for reviewing materials concerning rogue wave theories and tides; and John Odea and his colleagues at the U.S. Navy Carderock Division, Naval Surface Warfare Center, for reviewing materials concerning ship design. I thank Michael S. McDaniel of the law firm Countryman and McDaniel for permission to quote from their database of ship casualties at *www.cargolaw.com/presentations_casualties.html.*

Meaghan and Brad Van Liew kindly shared with me the behind-the-scenes effort to mount his "Around the World Alone" race and Brad's experiences during this challenging ordeal. I'm grateful to Peter Lewis, Brisbane, Australia, for providing background on the Sydney-Hobart races and for recounting his experiences in several races, including the 1998 race. Pete was always ready to take my e-mails and calls, in spite of the difference in time zones. Jen Hanson shared with me her recollections of the long night when the MV *Explorer* battled a North Pacific storm and the aftermath of being struck by a rogue wave that disabled the ship.

I gratefully acknowledge the cooperation and assistance of the staff at the Pacific Tsunami Warning Center, Ewa, Hawaii. I want to thank the National Oceanic and Atmospheric Administration and Center Director Dr. Charles McCreery for granting me access. I thank Marilyn Ramos for making arrangements, and a very special thanks to Barry Hirshorn and Stuart Weinstein for patiently answering my questions and for providing me with detailed insight into the operation of the center. Faye Wachs and Eugene Kim told me their incredible story of survival under the December 26, 2004, Southeast Asia tsunami.

The staff at the Balboa Branch of the Newport Beach Public Library (Phyllis Scheffler, librarian, assisted by Barbara Zinzer and Michael Payne), with its special collection of maritime literature, were wonderful to work with, as was the Nautical Museum at Newport Beach. I thank Kurt Mueller for providing the illustrations, Anne Elizabeth Powell for her usual insightful manuscript review, and Dr. Susanne Lehner for patiently responding to my questions and for writing the Foreword. As always, my wife Nancy, frequently first mate on my sailing ventures, freely gave her support, encouragement, proofreading, and research assistance.

Finally, to my agent, Ron Goldfarb, and to Jeff Robbins, Dick Morris, and the wonderful team at Joseph Henry Press, my gratitude.

Index

A

Adams, Karl, 197
Aebi, Tania, 182–183
Agulhas Current, 19, 79, 85, 118, 188–189, 203, 221, 239–240, 248
Aikau, Eddie, 174–175
Akiskalian, Eric, 121–123
Alaska Current, 79
Alexander Kielland, 206–207
America, 133
American Society of Naval Engineers, 233
Andaman Island. *See* Sumatra-Andaman Island earthquake of December 26, 2004
Anderson, Mark, 119–120
Anguilla, 199
Antarctica, 31–32
APL China, 220, 241
Arab traders, 18–19
Arabian Sea, 18, 59, 218
Arctic Ocean, 16
Argentina, 20
Arica, Chile, 133
Armada de Molucca, 19

Around the World Alone Race, 74–79, 189
Arthur M. Anderson, 204–205
Athletic shoes, as current indicators, 36–37
Atigun Pass, 202
Atlantic Ocean, 16–19, 35, 84–87, 97, 162
Australia Current, 27
Avalon Harbor, xx
Average depth of the oceans, 15
Azores, 96

B

Badger, 139–140
Baja California, Mexico, xix, 11, 17, 92, 242
Balance Bar, 75–78
Balboa Peninsula, xix, 228
Ballast, 124, 232–233
Ballast Point (Catalina Island), 124, 233
Baltic Sea, 18, 118
Banda Aceh, Indonesia, 152
Barker, Ernie, 138, 207–209
Barometer, 42–44, 84, 88, 91, 97–98

Barth, Joe, 201–202
Bascom, Willard, 3
Bass Strait, 25–29
Bay of Bengal, 18, 103–104, 150, 158
Beach, Edward L., Sr., 67
Beaufort, Sir Francis, 41
Beaufort Wind Scale, 41–42
Benjamin-Feir instability index, 240, 250
Bering Sea, 79
Bermuda High, 96
Bermuda Triangle, 8
Bernstein, Karl, 34
Blinsinger, Chase, 195–196
Board of Trade, certificate of
    seaworthiness, 188
Bobsled, 28
Borneo, 18, 218
Bradshaw, Ken, 116–117
Brady, Cornelius, 177–178
Breaking waves, 62–64
Bremen, 193, 197
Brindabella, 27–28
Broaching, 179
Brown, Emlyn, 190
Bryant, Edward, 133
Bulk carriers, 3–4, 53–54, 209–210,
    219–220, 223–225, 235–236
    high loss rate, 223, 245
Buoyancy force, 232
Burma Plate, 149, 152

C

Caledonian Star, 184–185,193
California, seismic design codes in, 228
California Current, 34
California Yacht Club, 124, 154, 233
Calm seas, 11–24
Canary Islands, tsunami potential, 162
Cape Horn, 75–77
Cape of Good Hope, 18–19, 26, 79, 85,
    188, 218
Cape Verde Islands, 35, 97, 200
Capillary waves, 12, 25, 38
Cargados Carajos Shoals, 126
Caribbean Sea, 16

Cargo ships, 220
Carl D. Bradley, 204
Carlsen, Henrik Kurt, 180–183
Carnival Cruise Lines, 221
Catalina Express, 71–72
Catalina Island, xviii, xx, 66, 71, 93–94,
    167, 175, 207, 233
Cat's paws, 38
Central America, 7
Challenger Deep, 20
Channel Islands, 34, 47, 61, 93, 124, 237
Chen, Henry, 241
Chile earthquake, May 22, 1960, 135–136
China, 9, 18
Christiansen, Bent, 137–138
Chubasco, 93
Clan McIntyre, 188
Clarke, Valerie, 71
Classification societies, 222
Cold fronts, 73–74
Columbus, Christopher, 19, 31, 89–90
Combination carriers (OBOs), 219
Coming Back Alive, 80
Composite waves, 49–54
    components of, 52
Confused seas, 88, 107, 111–112, 164–
    183, 192
    defining, 177
Constructive interference
    (superposition), 49–51, 207–210,
    249
Container ships, 3–4, 20–21, 36, 119,
    217, 220–221, 225, 228
Continental shelf, and shallow seas,
    205–207
Cook, Captain James, 25–26, 171
Cooper, Jessie B., 204
Coriolis force, 30, 33
Cortes Bank, xix, 116, 119
COSPAS-SARSAT network, 80
Crescent City, California, 136, 161
Crete, 130–131
Cross, Rear Adm. Bill, 86
Crozet Islands, 79, 189
Cussler, Clive, 190
Cycloids, inverted, 60
Cygnus Voyager, 6, 219

D

da Gama, Vasco, 19
Dampier, William, 32, 132
Daniel, Mann, Johnson and
 Mendenhall, 229
Davidson Current, 37
Davy Jones's locker, 226–246
Dead reckoning, 39
Deadweight tonnage, 217
Deep water, waves in, 14, 56, 61, 63, 98,
 115, 144, 203
Derbyshire, 53–54, 220, 222–225, 235,
 238, 245
 postmortem of the loss of, 222–225
Design
 of offshore structures, 230–231
 of port and coastal facilities, 227–
 230
 of ships, 216–219, 231–235, 238,
 246
Destructive interference, 49–50
Diaz, Bartolomeu, 19
Diffraction, 166
Dispersion, 61, 111, 165
Distinguished Flying Cross, 82
Dixon, Harold, 83–84
Doldrums, 31
Dominant period, 21
Dominican Republic, 67
Double-hulled tankers, 219, 221
Draft (or draught), 217, 241
Drag coefficient, 40
Draupner oil platform
 wave height on January 1, 1995, 192,
 209
 waves on, 207, 211
Dreams, 11, 16, 25, 47–48, 66–67, 71, 80,
 92, 164, 167, 173, 175, 187, 208,
 237, 240, 242–243
Drift, 34
Dwight D. Eisenhower (aircraft carrier),
 86–87

E

Earthquake Engineering Research
 Institute, 161
Earthquakes. *See also* Tsunamis; *and
 individual earthquakes*
 unpredictable as yet, 245
 unpreventable, 227
Earth's oceans and seas (Table 1), 16
East China Sea, 18
Ebbesmeyer, Curtis, 37
Edmund Fitzgerald, 204–205
Edrie, 139
Ekman transport, 33
Ekofisk North Sea oil field, 8, 206–207
Ellison, Larry, 27–28
Emerald Bay, 66
Emergency position indication radio
 beacon (EPIRB), 80–81
Energy, of waves, 5–6, 15, 23, 29, 41, 50,
 62, 67, 98, 111–113, 124–125, 163,
 167–169, 192, 202, 239, 249–250
EPIRB. *See* Emergency position
 indication radio beacon
Esprit d' Corp, 27–28
Eurasian Plate, 128
European Space Agency, 244
Evacuation warnings, 145
Evelina, 126
Extratropical cyclones, 74, 88
Extreme surfing, 116–117
Extreme waves, 6, 8, 14, 22, 41, 49, 54–
 56, 99, 116, 121, 145–146, 185,
 187, 189, 193, 199–200, 205–207,
 210–214, 221–222, 224, 227,
 230–231, 242, 245, 247–250
 evidence for, 212, 215
 predicting, 243–244
 probable causes, 247–248
 recent research on, 247–250
 riding, 118–121
Exxon Boston, 137–138
Exxon New Orleans, 137–138
Exxon North Slope, 137–138
Exxon Valdez disaster, 216

F

Fantome, tragedy of, 106–107
Fee, Jerry, 233
Fetch, 38–41, 88, 248
    maximum, 235
Flying Enterprise, 180–182
Foo, Mark, 117
Forecast accuracy, 105–106
    for hurricanes, 105–106
Fourier, Jean Baptiste Joseph, 22
Fourier analysis, 22
Fourier transform, 22, 49–50
Freak waves, 56, 193
Freedom America, 78–79
"furious fifties," 45
Future risk areas, 159–163

G

Gallimore, Skip, 194–197
Galveston, Texas, 104
Garrett, Chris, 248
Garrison, Keith, 207
Gioiello, Tony, 228–229
Glacier Bay National Park, 139–140
Global positioning satellite instrument,
    69
Global significant wave heights, 58
Gravity waves, 12, 38–39
Great Lakes, 203, 205
Greenland, 18
Grouper, 191–192
Grunion, 5
Guadalupe Island (Mexico), 92
Guam, 4, 20, 201, 218
Gulf of Alaska, 79, 137
Gulf of Mexico, 7, 73, 102, 104, 109,
    205, 230–231
Gulf Stream, 7, 191, 239–240, 248
Gulf War, 86
Gyres, 35

H

Hadley, George, 31
Hadley cells, 31
Hansa Carrier, 36
Hanson, Jen, 198
Harrison, Jon, 220, 241
Hawaiian Volcano Observatory seismic
    network, 143
Heaving to, 178, 180
Herodotus, 18
Highs, 42–44
Hilo, Hawaii, 9, 134–135
Hindcasts, 102, 213
Hirshorn, Barry, 141, 146–147, 158
HM Bark *Endeavor,* 26
Ho, Cheng, 18
Hogging, 190, 234–235
Hokkaido earthquake of July 12, 1993,
    230
Hokule'a, 173–175
Holdsworth, Ray, 84, 123
Homer, 6, 17, 38, 185
Horse latitudes, 31
Huntington Beach, xvii, 87, 114, 160
Hurricane Andrew, 96
"hurricane bow," 202
Hurricane Iniki, 196
Hurricane Katrina, 99, 104, 108, 149
Hurricane Luis, 200
Hurricane Mitch, 95, 106
Hurricane Roxanne, 102
Hurricane tracks, 105
Hurricanes, 7, 56, 69, 89–90, 95–96, 99–
    103, 107–109, 187, 200
    deadliest in the continental United
        States, 100–101
    forecasting, 99–103, 105–106
    a record season, 107–109
    storm surges and hurricane-induced
        flooding, 103–105
    typical life cycle, 97–99
    wind-wave patterns from, 96–97
    worst, 109
    See also Saffir-Simpson Hurricane
        Scale; Tropical cyclones;
        Typhoons

I

Iceland, 18
Idealized waves, 13–15, 48, 63
    not realistic, 21
    properties of, 14
Incorporated Research Institutions for
        Seismology (IRIS), 143
Indian Ocean, 17–18, 31, 79, 85, 91, 125,
        128, 147, 149, 151, 159, 188, 215,
        218, 228
Indian Plate, 149, 152
Indonesia, 9, 18, 128, 146, 215
Infragravity waves, 12
Interbox connectors (IBCs), 220
Intergovernmental Oceanographic
        Commission, 69, 141
International Association of
        Classification Societies (IACS),
        The, 223–224
    Recommendation 34, 224
International Convention on Load
        Lines, 238
International Maritime Organization, 23
Intertropical Convergence Zone, 31, 33
Irene, 180

J

Jannsen, Jan, 202–203
Japan, 9
Japanese tsunami, 132
Jason 1 satellite, 158
Jason and the Golden Fleece, 185
Java, 85, 140
Joint Tanker Project, 224
Joubert, Peter, 29
Junger, Sebastian, 194

K

Kaiwi Channel, 174
Kamsin wind, 32
Karess, xviii, 34
Kauai Channel, xi, 197

Kerguelen Islands, 79, 189
Kim, Gene, 151–157
King Cruiser, 152
Kingurra, 29
Kinsman, Blair, 22
Knots
    defining, 39
    wind and wave speed in, 39–40
Koh Phi Phi Island, 151–153, 158
    map of, 153
Kowloon Bridge, 54
Krakatoa, 126, 130, 140, 166, 215
Kuroshio Current, 53, 191, 239–240
Kuu Huapala, 197

L

La Conte, 80, 82
La Palma, 162
Labrador Current, 195
Lady Alice, 194–197
Laie Point, Oahu, 135
Lehner, Susanne, 213, 243–244, 249
Lewis, David, 171, 173
Lewis, Peter, 26–27, 29–30
Lightering, 218
Lightfoot, Gordon, 204
Lisbon, 133
Lituya Bay, 1, 79, 139, 146, 215
Lloyds of London, 222
    Casualty List, 3
    certificate of seaworthiness, 188
Load line, 236
Long Beach, 160
Los Angeles Times, 117
Loss rate, high for bulk carriers, 223, 245
Lows, 42–44
    notable, 44–45
Lyman-Mersereau, Marion, 174

M

M/V Explorer, 197–198
M/V MSC Carla, 4
MacArthur, Ellen, 45

Madagascar, 85, 151
Magellan (Fernão Magalhães), 19–20
Maintenance, of vessels, 225
Maldives, 151
Marina Del Rey, California, 11, 16, 34, 72–73, 154, 233
Marine, 7
Marine Electric, 235–239
    loss of, 235–239
Marine weather, forecasting and routing, 239–242
Maritime safety. *See* Ship safety
Maritime transport, evolution of, 217–222
Marx, Robert, 132
MaxWave project, 212–213
Mayday, 2, 28, 204
Mediterranean Sea, 16, 17–18
Metacentric height (GM), 232, 241
Michelangelo, 199
Midway Island, 173
Miller, Mike, 202
Minimum design, for wave heights, 225
Mistral, 32
Mobi (B&Q), 45–46
Modeling waves, 20–24
Modified Sieberg Tsunami Intensity Scale, 145
Monsoons, 59
Moore, Dan, 117, 119–120
Mount St. Helens, 163
Munk, Walter Heinrich, 22
Myles, Douglas, 131

N

Nailsea Meadow, 190
Napoleon, 22
National Hurricane Center, 97, 106
National Oceanic and Atmospheric Administration (NOAA), 37, 57, 61, 118, 241
    typical marine weather wind-wave forecast from, 57
National Research Council, 216
National Seismic Network, 143

National Weather Service, 97, 107, 109
    weather faxes from, 92, 194
Nazca Plate, 128
Nepenthe, 207–209
Neptune Sapphire, 188, 190–191
New Zealand to Tonga race disaster, 44–45
Newfoundland, 18, 58, 84, 91, 162, 193, 200
Newport, Rhode Island, 79
Newport Beach, California, 14, 62, 111
Newport Harbor, xvii, 93–94, 161, 169–170
Newport Harbor Nautical Museum, 94, 173
Newton's law of universal gravitation, 11
Nike shoes, 36
NOAA. *See* National Oceanic and Atmospheric Administration
North Atlantic, 84–87
North Atlantic Treaty Organization, 86
North Equatorial Current, 34–35
North Sea, 8, 54, 180–181, 187, 191, 206, 210, 215, 230–231
Northeast Trade Winds, 19, 31, 34

O

OBOs. *See* Combination carriers
Occluded front, 88
Ocean currents, 33–36
    strong, 188–192
Ocean exploration, history of, 18–19
Ocean surface winds, 33
Ocean Systems, Inc., 241
Ocean waves, types of, 13
Oceans. *See individual oceans*
Oceanweather, Inc., 242
Odysseus, 6–7, 17, 185
Odyssey, The , 6, 17, 38
Offshore structure design, 230–231
Oil platforms, 29
Oistad, Erik, 242
Okhotsk Sea, 16
Old Testament tsunamis, 131

Orbital motion, 192
Osthav, 180

P

Pacific High, 30, 42, 97
Pacific Ocean, 16–17, 25, 32, 36–37, 42,
    97, 111, 124, 127–128, 141–145,
    171–175, 218, 241
Pacific Tsunami Warning Center, 142,
    146–147, 149, 158–159
Palos Verde tsunami (scenario), 161
Pampero, 32
Panamax, 217
Papua New Guinea tsunami, 147, 160
Paragon, 94
Parametric rolling, 221
Pararas-Carayannis, George, 69
Paxton, George, 197
Perfect Storm, The , 194
Periodic function, 22
Periods, 12–15, 23, 34, 49–50, 56, 62–64,
    69, 96, 118, 125, 138, 212, 240,
    250
    dominant, 21
    return, 210
Persian Gulf, 18–19, 86, 218
Philippine Sea, 20
Phoenicians, 18
Phuket Island, Thailand, 151–153
Piailug, Mau, 171, 173–174
Pine Ridge, 225
Pitchpoling, 63, 179
Pittsburgh, 201
Plimsoll marks, 236
Polynesian navigation, by wave
    patterns, 170–175
Port of Los Angeles, 71, 137, 209, 220,
    228–229
Port Royal, Jamaica, 132
Portuguese navigators, 19, 162, 188
Poseidon, 7, 17, 227
Prince William Sound earthquake, 136
Probability, of large waves, 54–59
Punta del Este, 77–78

Q

Qiantang River, 65–66
Queen Elizabeth, 194
Queen Elizabeth 2, 109, 185–186, 200,
    215, 221
Queen Mary, 193–194

R

Rayleigh distribution, 55, 249
    of wave heights, 55
Recurving, storm paths, 96–97, 108
Red Sea, 18–19
Reedy, Mark, 71
Reef break, 115
Reflection, of waves, 61, 166–169
Refraction, of waves, 61, 166–169
Remijan, Mark, 220, 241
Rescue Coordination Center, 81
Rescues, 28–29, 44, 79–82, 102–103, 122
Resonant effect, 40
Return period, 210
Rhodius, Apollonius, 185
Rift zones, 162–163
Ring of Fire, 127–128, 141
"roaring forties," 45
Rogue Wave, 209
Rogue waves, 4, 8, 56, 81, 109, 115, 178,
    181, 186, 188, 191–192, 195–199,
    202, 209, 213, 234, 237, 244–245,
    249
    causes and characteristics of, 185–
    187
Rolling, parametric, 221
Royal Caribbean Cruises Ltd., 221
Ruau-Moko, 127, 227

S

Safety. *See* Ship safety
Saffir-Simpson Hurricane Scale, 95–96
Sagging, 234–235
Salerno, Ugo, 224
San Andreas Fault zone, 159
San Clemente Island, 92–93

San Diego, 114

San Pedro, California, 160

San Pedro Channel, 66, 72, 187

Sansinena II, 137–138

Santa Ana winds, 32, 114–115

Santa Barbara Island, 34, 208

Santa Monica Bay, xviii, 34

Santa Rosa Island, 47, 237

Santo Domingo Harbor, 67

Santorini Islands, 130

Satellites, 58, 76, 80, 89, 97, 102, 106–107, 125, 144, 158, 176, 196, 213, 220, 227, 240–241, 243–244

Sayonara, 27–28

Scantlings, 234

"screaming sixties," 45

Scotch Cap lighthouse, 9, 134–135

Scripps Institution of Oceanography, 22, 118

Sea anchors, 179

Sea clutter, 243

Sea of Cortez, 17, 92

Sea of Japan, 18, 139

Sea severity, 109

Seal Beach, 160

Seas, 7, 12, 18, 21, 23, 27, 29, 35–36, 40, 45–47, 53–54, 58, 62, 72–73, 79, 81–82, 85–86, 88, 150, 165, 169, 179–181, 189, 194–195, 201–202, 204, 221, 232, 237–238, 248
    calm, 11–24
    storm-tossed, 71–109

Seascape, 125–126

Seiches, 65, 144

Seismic design codes, 228

Seismic sea waves, 128–129

Seismometers, 23

Semester at Sea Program, 197–198

Sets
    of currents, 35, 171
    of waves, 15, 56, 61, 111, 179

Shallow seas, and the continental shelf, 205–207

Shallow water, waves in, 14, 56, 115, 203–204

Shema, Rick, 242

Ship design, 216–219, 231–235

"Ship of Gold," 7–8

Ship safety, 216–225
    crews being sent to sea to die needlessly, 245–246
    increasing, 223

Shipwrecks, reasons for, 2–4

Shoaling, 61, 72

Sidereal compass, 172

Significant wave heights, 21, 99, 118, 210, 235
    global, 58

Sine waves, 13, 22, 49–51

Single-hulled tankers, 225

Sirocco, 32

Sitting Tall, xx

Sixty Mile Bank, 11, 16, 24

Small craft advisory, 42

Smith, Nancy, 187

Snow, George, 27

Solitary waves, 176

Solitons, 176

SOS. *See* Mayday

South China Sea, 18

South Equatorial Current, 34–35

Southeast Asia tsunami. *See* Sumatra-Andaman Island earthquake of December 26, 2004

Southern Ocean, 26, 59, 77–78, 184, 188–189, 193

Spanish Armada, 226

Spanish navigators, 7, 19, 89, 162

Spatial focusing, 249

Spencer, Russell, 242

Spice Islands, 19–20

Squalls, 82–84

Sri Lanka, 150

SS *Lane Victory*, 138, 209

SS *Pennsylvania*, 177–178

SS *United States*, 6

SS *Waratah*, 188–190, 232, 245

St. Barthélémy, 199

St. Martin Island, 199

Stand Aside, 28–29

Stars, zenith, 173

Steepness, of waves, 63–64

Stokes, George G., 21
Storm surges, and hurricane-induced
    flooding, 103–105
Storms, 89–95, 192–205
    origins of, 73–74
Straits of Gibraltar, 18
Subduction, 149
Subic Bay, 85
Suez Canal, 217–218
Sumatra, 8, 18, 85, 128, 140, 146, 148–
    152, 158–159, 176, 218
Sumatra-Andaman Islands earthquake
    of December 26, 2004, 8–9, 128,
    147–163, 176, 226
Sunda Straits, 85, 140, 152
Sunda trench, 152
Sunmore, 139
Superposition, 49–51, 207–210, 249
Surfing, 110, 112–115
    extreme, 116–117
Survival strategies, 178–180
Sverdrup, Harald Ulrik, 22
Swells, 12, 110–126, 170
    defining, 111–112
    large, 112–113
    long-period travelers, 123–124
    in midocean, 124–126
    tracking, 117–118
    see also Surfing
Sword of Orion, 28
Sydney-Hobart race, 25–29
Synthetic aperture, 243

T

Tankers, 3–4, 137, 202, 209–210, 216–
    219, 224–225, 232, 236
    defective, 224
    double-hulled, 219, 221
    single-hulled, 225
Tasman Sea, 26, 29, 209
Teal Arrow, 200
Tempests, 71–109
Temporal focusing, 249
TEUs (twenty-foot equivalent units),
    217

Texas Tower 4, 206
Thailand, 9, 18, 151–152
Theodora, 237
Thira (Thera) Island, 130
    map of, 131
Thomas, Steve, 171
Thompson, Tommy, 7
"Three Sisters," 69, 186, 205, 209
Thunderstorms, 87
TI Europe, 218–219
Tidal bores, 65–66
Tidal streams, 32, 176
Tides, the pull of the moon, 64–65
Time history, 23
Titanic, xviii, 193
Torricelli, Evangelista, 42
Towsurfing, 121–123
Trade winds, 32–33
Transtidal waves, 12
Treasure Fleet, Chinese, 18
Tropic of Cancer, 74
Tropical cyclones, 89–90, 240
Tropical depressions, 44, 53, 69, 89–90,
    92, 97
Tsunami bulletins, 141–142
Tsunami Intensity Scale, 145–147
Tsunamis, 6, 8–9, 15, 63, 69, 127–147,
    149–151, 154, 163, 176, 226–227
    in ancient times, 130–132
    caused by landslides and volcanoes,
        139–140
    characteristics of, 128–130
    diving under a killer, 151–159
    education programs, 244
    historically significant, 132–133
    Japanese, 132
    Old Testament, 131
    recent examples, 133–139
    the "Ring of Fire," 127–128
    travel time in hours, in the Chile
        earthquake, May 22, 1960, 136
    tsunami intensity scale, 145–147
    walls, 230
    warning systems, 140–145, 227, 245
    wave height and run-up, 129
Turmoil, 181–182

Tyne Bridge, 53
Typhoon Orchid, 53, 222, 235
Typhoon Paka, 4
Typhoons, 4, 53, 89, 90, 192, 200–201,
    222

U

Ulrich, Howard, 139
Ultra large crude carriers (ULCCs), 218
Ultragravity waves, 12
Unimak Island, Alaska, 9, 134, 135
United Nations Educational, Scientific,
    and Cultural Organization, 69,
    141
U.S. Air Force Reserve, 106
U.S. Army Corps of Engineers, 228
U.S. Coast Guard, 9, 80–81, 237
U.S. Department of Commerce, 57
U.S. Geological Survey (USGS), 143
U.S. Naval Academy, 233
U.S. Navy, 99
USS *Castine*, 68
USS *Forrestal*, 84–85, 201
USS *Hancock*, 201
USS *Memphis*, 67–70
USS *Ramapo*, 211–212, 215
    wave height measurement scheme
        on, 214
USS *Taylor*, 233

V

Valley Forge, 202
Van Liew, Brad, 74–75, 79, 189
Van Liew, Meaghan, 74, 78
Velella velella, 5
Very large crude carriers (VLCC), 218
Vessel Optimization and Safety System
    (VOSS), 241–242
Victoria, 20
Vikings, 18
Volcanic calderas, 162

W

Wachs, Faye, 151–157
Walker, Spike, 80
"WAM" site, 118
Warantah, 188–190
Warm fronts, 73
Warning systems, tsunami, 140–145,
    227, 245
Warps, 179
Warwick, Peter, 185–186, 200
Wateree, 133
Waterspouts, 87
Watkins, Bill, 124
Wave addition, 49
Wave amplitude, 13, 75
Wave buildup, near shore, 66–70
Wave diffraction, 166, 169–170
    at the Newport Harbor entrance,
        171
Wave forecasting algorithms, 243
Wave heights, 13–15, 21–22, 39–41, 50–
        51, 54–59, 62, 75–76, 99, 106,
        118, 125, 129, 135–136, 140–141,
        160, 165, 185–186, 196–197, 201,
        210–213, 217, 224–225, 231, 244,
        250. *See also* significant wave
        heights
    minimum design for, 225
    versus wind speed, duration, and
        fetch, 41
Wave interactions, 166–167
Wave interference, destructive and
        constructive, 49–51
Wave modeling, 20–24
Wave movement, 59–61
Wave patterns, Polynesian navigation
        by, 170–175
Wave properties, 13–16
Wave reflection, 167
Wave refraction, 167–168
Wave speed, 14, 62, 167
Wave steepness, 15, 64, 244
Wave types, 12–13

Wavelength, 13–15, 49, 62–63, 125, 128–129, 167, 188, 192, 203, 213, 221, 235, 244
Weather forecasts, 105–106
Weather Reconnaissance Squadron, 106
Weather routers, 242
"The Wedge" (surfing spot), 114
Weinstein, Stuart, 141
Wessel, Paul, 145
White, Linda, 125
White, Tod, 125
Wind speed, 39
Wind waves, 37–39
    growth of, 40–41
    patterns from a hurricane, 98
    theory of, 23
Wind Waves, 22

Winston Churchill, 29
World Glory, 188, 190
World merchant fleet, 217–218
    distribution of vessels in, 218
"The Wreck of the Edmund Fitzgerald," 204

Y

Young, Samuel, 104–105
Youngquist, Andy, xx

Z

Zenith stars, 173